D1066765

BASIC GAS CHROMATOGRAPHY

by

H. M. McNair

&

E. J. Bonelli

Lithographed by Consolidated Printers, Berkeley, California

Printed in the U.S.A.

1st Printing—February, 1965
2nd Printing—April, 1966
3rd Printing—February, 1967
4th Printing—January, 1968

This book is available only from Varian Aerograph and their official subsidiaries and representatives. U.S. Price $2.50 paperback; $5.00 hardback. Prices slightly higher in other countries due to transportation and duty charges.

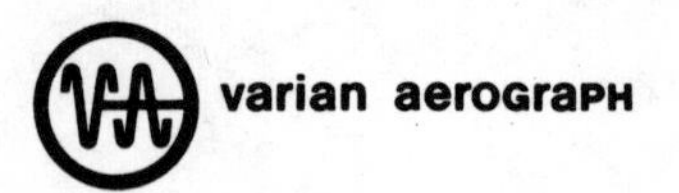

2700 Mitchell Drive
Walnut Creek, California 94598
U.S.A.

Pelikanweg 2
4002 Basel, Switzerland

PREFACE

The primary objective of this book is to select the basic aspects of the G. C. technique and make them available in a simple and easy to understand manner. An effort has been made to present useful data such as calibration factors for detectors, chemical structure of liquid phases; methods for preparing liquid phases and packing columns. No attempt has been made to completely cover the subject but rather to provide a basis for understanding and selecting references so that the eager student can continue his study. We feel that a major advantage of G. C. is its simplicity; hopefully this book will confirm that feeling and encourage more people to use the technique. Another objective was to keep the price low so that students throughout the world could purchase the book.

This book is a result of the Basic Gas Chromatographic Courses which Varian Aerograph offers throughout the world. The first edition in 1965 was a collection of lectures which Aerograph people had been presenting. The second edition in 1966 was expanded to include more literature references and more reference material such as structure of liquid phases and maximum temperature limits of liquid phases. Because of requests, it was made available to the public at large. The demand has been so high that we have prepared a third edition and printed it in both hardback and paperback editions. In this new edition, we have expanded the chapter on theory and the chapter on detectors; and rewritten the quantitative analysis chapter to include statistical treatment of data and more information on integration techniques. In addition, we have added the laboratory exercises which should make the book attractive for use in colleges and universities.

We would like to acknowledge the assistance and efforts of our colleagues at Varian Aerograph. They have provided useful information and suggestions. Most of the experimental work was conducted in the research and applications laboratories of Varian Aerograph. In addition, we would like to thank the students of the Aerograph Training Courses whose comments and suggestions have encouraged us to make this edition available.

H. M. M.
E. J. B.

Walnut Creek, California
February, 1967

DEDICATION

This book is dedicated to our friends and colleagues at Varian Aerograph. It is both a pleasure and a stimulus to work with them.

AUTHORS

Harold M. McNair obtained his B.S. Chemistry, magna cum laude from the University of Arizona in 1955. Purdue University granted him a M. S. in 1957 and Ph. D. in 1959 for theses in coulometric titrations and gas chromatography. Awarded a Fulbright post doctorate fellowship, he studied under Prof. A. I. M. Keulemans at Eindhoven, The Netherlands. He met and married his wife, Marijke, while in Holland. They returned to the United States in October, 1960, and he worked as a chemist for Esso Research and Engineering, Linden, New Jersey. He joined F & M Scientific in 1962 and opened their European Subsidiary in Amsterdam. He served as European Sales Manager, Technical Director, and General Manager before joining Wilkens Instrument in 1964 as Director of International Operations. He is presently Marketing Manager for Varian Aerograph. He has published six papers and two chapters on Gas Chromatography.

Ernest J. Bonelli obtained his B. S. in Chemistry from the University of San Francisco in 1958 and his M.S. from the University of the Pacific in Stockton. He was employed for several years as a research chemist at United Vintners, Inc., in Asti, California. Mr. Bonelli joined Varian Aerograph in September, 1962, as a chemist in Research and Development, working in pesticide analysis with the electron capture detector. This work led to the publication of the "Pesticide Residue Analysis Handbook". He has also authored papers in the gas chromatographic analysis of steroids, fusel oil, and lead alkyls. Mr. Bonelli has also authored two chapters dealing with pesticides and dual channel gas chromatography. He is presently Manager of Applications Department, and in charge of the Training courses and Symposia sponsored by Varian Aerograph.

TABLE OF CONTENTS

I. INTRODUCTION

A. DEFINITION

The basis for gas chromatographic separation is the distribution of a sample between two phases. One of these phases is a stationary bed of large surface area, and the other phase is a gas which percolates through the stationary bed.

Gas Chromatography is a technique for separating volatile substances by percolating a gas stream over a stationary phase. If the stationary phase is a solid, we speak of *Gas-Solid Chromatography* (G. S. C.). This depends upon the adsorptive properties of the column packing to separate samples, primarily gases. Common packings used are silica gel, molecular sieve, and charcoal.

If the stationary phase is a liquid, we speak of *Gas-Liquid Chromatography* (G. L. C.). The liquid is spread as a thin film over an inert solid and the basis for separation is the partitioning of the sample in and out of this liquid film. The wide range of liquid phases with usable temperatures up to 400°C make G. L. C. the most versatile and selective form of gas chromatography. It is used to analyze gases, liquids, and solids. Most of our discussion will be concerned with G. L. C.

B. HISTORY

Chromatography was first employed by Ramsey[1] in 1905 to separate mixtures of gases and vapors. These first experiments used selective adsorption on, or desorption from, solid adsorbents such as active charcoal. The following year Tswett[2] obtained discrete colored bands of plant pigments on a chromatographic column. He coined the term "chromatography", (literally: "Color writing") which is obviously a misnomer when applied to current methods.

Following the suggestion of Martin and Synge[3] in a study for which they were later awarded the Nobel Prize, James and Martin introduced gas-liquid chromatography in 1952[4, 5]. The sensitivity, speed, accuracy, and simplicity of this method for the separation, identification, and determination of volatile compounds has resulted in a phenomenal growth. Presently there are more than 14,000 G.C. references, and the growth rate is about 1800 - 2000 per year. It is estimated that 50,000 gas chromatographs are in use.

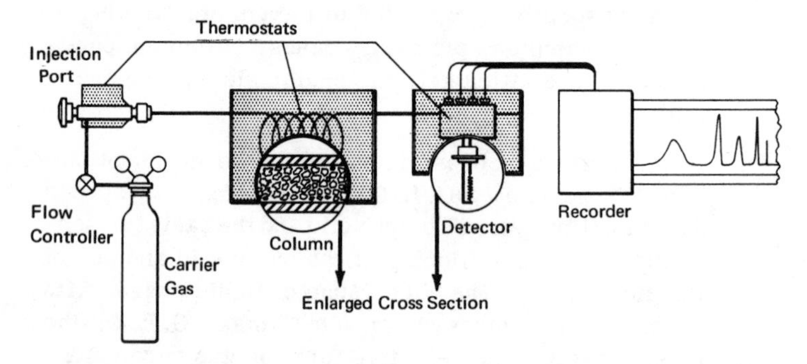

FIGURE I-1—SCHEMATIC DRAWING OF A GAS CHROMATOGRAPHIC SYSTEM

C. APPARATUS

The basic parts of a gas chromatograph are:

1. Cylinder of carrier gas
2. Flow controller

3. Injection port (sample inlet)
4. Column
5. Detector (with necessary electronics)
6. Recorder
7. Thermostats for injector, column, and detector.

These are discussed in detail in Chapter II.

D. TECHNIQUE

In G. L. C. the components to be separated are carried through the column by an inert gas *(Carrier Gas)*. The sample mixture is partitioned between the carrier gas and a non volatile solvent *(Stationary Phase)* supported on an inert size-graded solid *(Solid Support)*. The solvent selectively retards the sample components, according to their distribution coefficient, until they form separate bands in the carrier gas. These component bands leave the column in the gas stream and are recorded as a function of time by a detector.

Advantages of this elution technique are:

1. The column is continuously regenerated by the inert gas phase.
2. Usually the sample components are completely separated and mixed only with an inert gas making collection and quantitative determinations easy.
3. The analysis time is short.

A disadvantage is that strongly retained components travel very slowly, or in some cases do not move at all. This difficulty can be overcome by using temperature programming of the column to decrease elution time. *Temperature Programming* is the increase of column temperature during an analysis to provide a faster and more versatile analysis. This technique is discussed in detail in Chapter IX.

E. RESULTS

When using a strip chart recorder, the written record obtained from a chromatographic analysis is called a *Chromatogram.* Usually time is the abscissa and millivolts the ordinate. A chromatogram illustrating the results one can obtain is shown in Figure I-2.

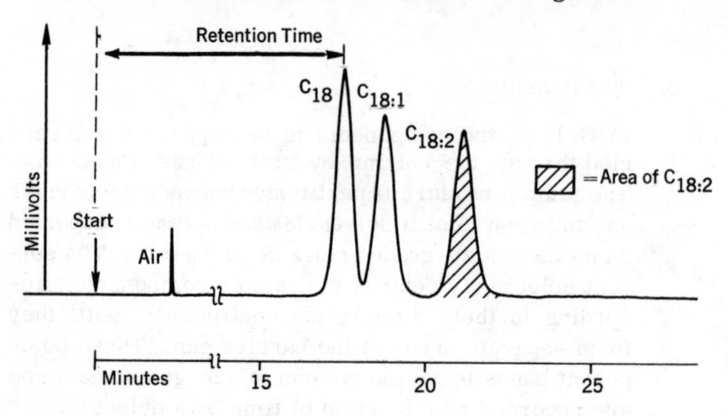

FIGURE I-2—TYPICAL CHROMATOGRAM; FATTY ACID ESTERS

F. ADVANTAGES OF GAS CHROMATOGRAPHY

Specific reference is made to Figure I-2 with comments applicable to the technique in general.

1. Speed

   The entire analysis is completed within 23 minutes. The use of gas as the moving phase has the advantage of rapid equilibrium between the moving and stationary phases and allows high carrier gas velocities to be employed. Separations requiring only seconds have been reported, however, analysis time of minutes duration is more common in G.L.C. Preparative scale separations, or resolution of wide-boiling complex samples may require hours.

2. Resolution

Peaks C_{18}, $C_{18:1}$, and $C_{18:2}$ represent the methyl esters of stearic, oleic, and linoleic acids. The separation of these compounds by other techniques is extremely difficult or impossible. The boiling point difference is negligible in that the compounds vary only in the degree of unsaturation. Gas chromatography can separate compounds with almost identical boiling points. By using selective solvents G. C. provides resolution impossible by distillation or other techniques.

3. Qualitative analysis

The *Retention Timc* is that time from injection to the peak maxima. This property is characteristic of the sample and the liquid phase at a given temperature. With proper flow and temperature control, it can be reproduced to within 1% and used to identify each peak. Several compounds can have identical or close retention times, but each compound has only one retention time. This retention time is not influenced by the presence of other components.

4. Quantitative analysis

The area produced for each peak is proportional to that peak's concentration. This can be used to determine the exact concentration of each component. In Figure I-2, the peak areas are 36.7%, 33.0%, and 30.3% (relative area as measured by disc integrator), whereas the actual concentrations were 36.4%, 33.2%, and 30.4% for stearic, oleic, and linoleic methyl esters. Accuracy attainable with G. C. depends upon technique, detector, integration method, and sample concentration. Chapter VII discusses these points in detail. For major peaks with manual area measurements, one can expect accuracies of 1 to 2% relative, or better.

5. <u>Sensitivity</u>

A major reason for the extensive analytical application of G.C. is the sensitivity available. The simplest forms of thermal conductivity cells can determine down to 0.01% (100 ppm). The flame detector easily sees parts per million, and the specific electron capture and phosphorus detectors can measure parts per billion or picograms (10^{-12} g). An added advantage of this extreme sensitivity is the small sample size required. Microliters of sample are sufficient for complete analyses.

To gain some appreciation of this extreme sensitivity, consider a component having a molecular weight of 100 which is eluted in a peak of 10 seconds duration (width at the baseline). If carrier gas flow is 30 milliliters per minute, the eluted component will be contained in a 5 milliliter volume of carrier gas (10 sec. x 30 ml/min.). Thus, 10^{-14} moles of component (10^{-12}gm/100 gm per mole) is contained in approximately 2×10^{-4} moles of carrier gas (5 ml/22,400 ml per mole). The ratio $2 \times 10^{-4}/10^{-14}$ means that there are 20 billion (2×10^{10}) molecules of carrier gas for each molecule of component in the eluted peak. This is indeed trace analysis.

G. USES OF GAS CHROMATOGRAPHY

1. <u>Analytical methods</u>

G. C. is used to identify and determine any material having an appreciable vapor pressure (1 to 1000 mm) at column operating temperature (-70°C to +400°C). Many solids have been analyzed by their characteristic "cracking" patterns formed at even higher temperatures. In view of the advantages mentioned above, it is often called *The poor man's mass spectrometer.* It takes complex samples, often too difficult for other techniques, separates them into individual components and allows both qualitative and quantitative analysis in minutes.

2. Determination of physical properties

   Many physical properties, such as surface area, adsorption isotherms, heats of solution, activity coefficients, partition coefficients, molecular weight, and vapor pressure can be readily measured by G. C. techniques. These procedures usually yield results in minutes comparable in accuracy to the slower classical methods.

3. Preparative methods

   G. C. can be used for isolating pure components on a gram basis. Instruments are now available for the automatic injection, separation, and collection of samples.

BIBLIOGRAPHY

1. *Ramsey, W., Proc. Roy. Soc. A76 111 (1905).*
2. *Tswett, M., Ber. deut. botan. Ges. 24, 316, 384 (1906).*
3. *Martin, A.J.P. and Synge, R.L.M., Biochem J. 35, 1358-1368 (1941).*
4. *James, A.T. and Martin, A.J.P., Biochem J. (Proc.) 48, vii (1951).*
5. *James, A.T. and Martin, A.J.P., Analyst 77, 915-932 (1952).*

II. CHROMATOGRAPHIC SYSTEM

A. CARRIER GAS

A high pressure gas cylinder serves as the source of carrier gas. In isothermal G.C. the permeability of a column does not change during an analysis. A pressure regulator is used to assure a uniform pressure to the column inlet, and thereby a constant rate of gas flow. At a given temperature, this constant rate of flow will elute components at a characteristic time (the retention time). Since the flow-rate is constant, the components also have a characteristic volume of carrier gas (the retention volume).

Commonly used gases are hydrogen, helium, and nitrogen. The carrier gas should be:

1. Inert to avoid interaction with the sample or solvent.
2. Able to minimize gaseous diffusion
3. Readily available, pure
4. Inexpensive
5. Suitable for detector used

Column efficiency depends upon choosing the proper linear gas velocity. A common value for 1/4" O. D. columns is 75 ml/minute; for 1/8" O.D. columns, 25 ml/minute. The optimum flow rate can be easily determined experimentally by making a simple Van Deemter plot of HETP vs. linear gas velocity (see Figure II-1). The most efficient flow-rate is at the minimum of HETP or maximum plates. Consult Chapter III for additional details.

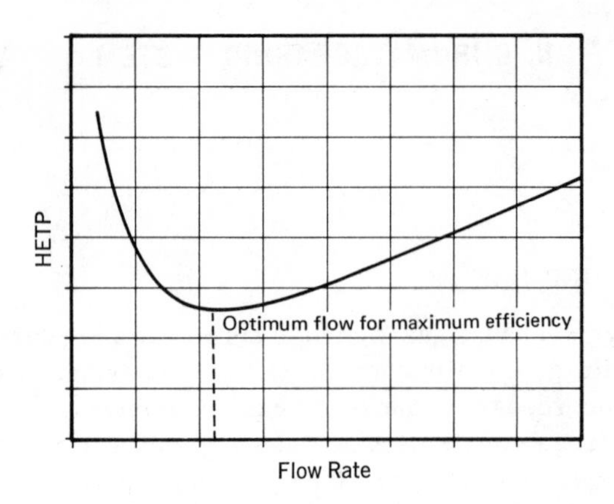

FIGURE II-1—van DEEMTER PLOT

The simplest way to measure gas flow rates is with a soap-bubble flowmeter and a stopwatch. See Appendix for details of this procedure.

B. SAMPLE INTRODUCTION

The sample should be introduced instantaneously as a "plug" onto the column. A good check on the sampling technique is to raise the injection heater temperature and reduce the sample size. If either of these factors increases the number of theoretical plates, a poor sampling procedure was being used.

Gases are usually introduced by gas-tight syringes (Figure II-2) or by-pass sample loops (Figure II-3). Reproducibility is better than 0.5% relative with sample loops.

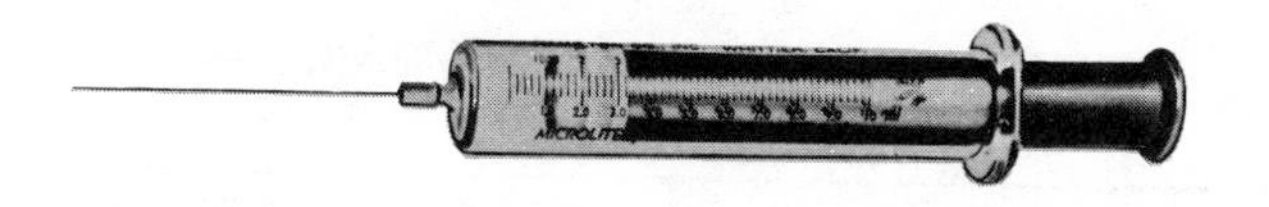

FIGURE II-2—10.0 ml GAS TYPE SYRINGE

FIGURE II-3—GAS SAMPLING VALVE

Liquids are handled with syringes. Recently devices for the direct injection of solids have become commercially available. The easiest technique however for solids is solution in a solvent whose response does not interfere with the samples being analyzed.

A standard technique for the introduction of gases and liquids is to introduce a hypodermic syringe needle through a self-sealing serum cap and inject measured volumes from an attached syringe. These syringes are commercially available in various volumes down to one microliter and with proper handling can give a reproducibility of 2% relative. Gas-tight syringes are also available.

TABLE II-1—SAMPLE VOLUMES FOR DIFFERENT COLUMNS

COLUMN TYPE	SAMPLE SIZES GAS	LIQUID
Preparative 1" O. D., 20% Liquid	0.05 - 5 liter	0.02 - 2 ml
Regular Analytical 1/4" O. D., 10% Liquid	0.5 - 50 ml	0.2 - 20 μl
High Efficiency 1/8" O. D., 2% Liquid	.1 - 1 ml	0.04 - 4 μl *
Capillary 1/16" O. D., 5.0 u film	0.1 - 10 ul	0.004 - 0.5 μl *

*These sample sizes are often obtained by sample splitting techniques.

C. COLUMN

The column tubing can be made from copper, stainless steel, aluminum, and glass in a straight, bent, or coiled form. Copper may be unsuitable in that it shows absorption or reaction with certain sample components (amines, acetylenes, terpenes and steroids).

In general, stainless steel columns are used; packed while straight to obtain a uniform packing and coiled to facilitate long lengths. Straight columns are more efficient, but can be cumbersome, particularly for work at high temperatures. If coiled, the spiral diameter should be at least ten times the column diameter to avoid diffusion and racetrack effects.

Packed column lengths vary from a few inches to more than 50 feet in length. Common analytical columns are 3 to 10 feet in length. Longer lengths give more theoretical plates and resolution. The carrier gas velocity changes during passage through the column and thus only a short section of the column operates at the optimum flow rate. This means that with extremely long columns, the plates and resolution obtained show diminishing returns. In addition, long columns require very high inlet pressures. High pressures present problems in injection technique and avoiding gas leaks. An advantage of long columns, however, is that sample capacity is proportional to the amount of liquid phase present. This means that larger sample sizes may be injected on longer columns. See Table II-1 for typical sample sizes.

Column diameters vary from 0.01 up to 2 inches I. D. and larger. The smaller the column diameter the higher the column efficiency. Standard analytical columns are 1/8 and 1/4 inch O. D.; Capillary and Micro Pak® columns for a large number of theoretical plates are 1/16 inch O.D. An obvious way to increase the column capacity for sample is to increase column diameter. Preparative scale separations are run on 3/8 inch, 1/2 inch, and larger diameter columns. Unfortunately, poor diffusion and the multi-path effect (see Chapter III) cause a decrease in column efficiency with increased column diameter.

D. SOLID SUPPORT

The purpose of the solid support is to provide a large uniform, inert surface area for distributing the liquid phase. Some of the desirable support properties are:

1. Inert (avoid adsorption)
2. High crushing strength
3. Large surface area
4. Regular shape, uniform size

There are two basic Chromosorb* products used in G.C. - Chromosorb P and Chromosorb W. Chromosorb P (pink) grades are prepared from Johns Manville C-22 firebrick, while the Chromosorb W (white) grades are prepared from Johns Manville celite filter aids. Chromosorb P is used where the highest column efficiency is desired. The surface of this material does, however, show strong adsorption of polar compounds. It is a calcined material, pink in color and relatively hard. Compared to Chromosorb P, Chromosorb W is used where a relatively inert surface is required. The column efficiency obtained is not as high as that obtained with Chromosorb P. It is a flux calcined material, white in color and relatively soft. It is more inert than P and is recommended for polar compounds. Additional details are found in Chapter IV.

E. STATIONARY PHASE

The correct choice of the partitioning solvent to be used is probably the most important parameter in G.L.C. Ideally the solvent should have the following characteristics:

1. Samples must exhibit different distribution coefficients.
2. Samples should have a reasonable solubility in the solvent.
3. Solvent should have a negligible vapor pressure at operating temperature.

The versatility and selectivity in G.L.C. is due to the wide choice of solvents available. The partition coefficient ratio (relative volatility above the solvent), can change fifty-fold in different liquids. This results in fifty-fold differences in retention times and makes separation easy. Choosing the proper stationary phase

* *Chromosorb is Johns Manville registered trademark for support material for G.C.*

is an extremely important task. The chapter on columns provides additional information.

F. TEMPERATURE

To be exact, one must describe the temperature of the injection chamber, the column, and the detector. Since the temperature of all three of these component parts serves different functions, it is desirable that the instrument possesses three different temperature controls.

1. Injection-port temperature

   The injection port should be hot enough to vaporize the sample so rapidly that no loss in efficiency results from the injection technique. On the other hand, the injection-port temperature must be low enough so that thermal decomposition or rearrangement is avoided. A practical test is to raise the temperature of the injection port. If the column efficiency or peak shape improves, the injection port temperature was too low. If the retention time, the peak area or the shape changes drastically, the temperature may be too high and decomposition or rearrangement may have occurred.

2. Column temperature

   The column temperature should be high enough so that the analysis is accomplished in a reasonable length of time, and low enough that the desired separation is obtained. According to a simple approximation made by Giddings,[1] the retention time doubles for every 30° decrease in column temperature. For most samples the lower the column operating temperature, the higher the ratio of partition coefficients in the stationary phase and the better the resultant separation. In some cases it is not possible to use a low operating temperature; and particularly in the case of wide boiling samples, it will be desirable to employ temperature programming.

3. Detector temperature

The influence of temperature on the detector depends considerably upon the type of detector employed. As a general rule, however, it can be said that the detector and the connections from the column exit to detector must be hot enough so that condensation of the sample and/or liquid phase does not occur. Peak broadening and loss of component peaks are characteristic of condensation. The stability and resultant usable sensitivity of a thermal conductivity detector depends upon the stability of the detector temperature control. It should be $\pm$ 0.1^{o}C. For ionization-type detectors, temperature control is not as critical. The temperature must be maintained high enough to avoid not only condensation of the samples, but also of the water or by-products formed in the ionization process.

G. DETECTORS

The detector indicates the presence and measures the amount of components in the column effluent. Desirable characteristics of a detector are high sensitivity, low noise level, a wide linearity of response, response to all types of compounds, rugged, insensitive to flow and temperature changes, and inexpensive. There is no ideal detector; however, the thermal conductivity cell and flame ionization detector come close to being universal detectors. In addition, specific detectors such as the electron capture and phosphorus detector have the advantage in that they selectively detect only certain types of compounds. This makes them extremely useful for quantitative and qualitative analysis.

The two most popular detectors are thermal conductivity cell (TC) and flame ionization detector (FID). The TC detector employs a Tungsten filament which is heated by passing constant current through it. Carrier gas flows continuously over this heated filament and dissipates heat at a constant rate. When sample molecules mixed with carrier gas pass over the hot filament, the rate of heat loss is reduced and the resistance of the filament increases. This resistance change is easily measured by a Wheatstone bridge and the signal fed to a recorder where it appears as a peak. The principle of operation is that the ability to conduct heat from a filament is a function of the molecular weight of the gas.

In the FID detector, hydrogen and air are used to produce a flame. A collector electrode with a DC potential applied is placed above the flame and measures the conductivity of the flame. With pure hydrogen, the conductivity is quite low; however, as organic compounds are combusted, the conductivity increases and the current which flows can be amplified and fed to a recorder.

More detail on these and other types of detectors is discussed in Chapter V.

H. RECORDER

Present practice is to use a strip chart recorder to obtain a permanent record of the results. A 1 mv, 1 sec. full scale response is recommended.

The potentiometric type recorder used in G.C. is a servo-operated voltage balancing device (Figure II-4).

It operates on the principle that the input signal (V_1) is continuously balanced by an equal and opposite polarity feedback signal (V_2). Simultaneously, the chart is driven in a direction at right angles to the pen motion.

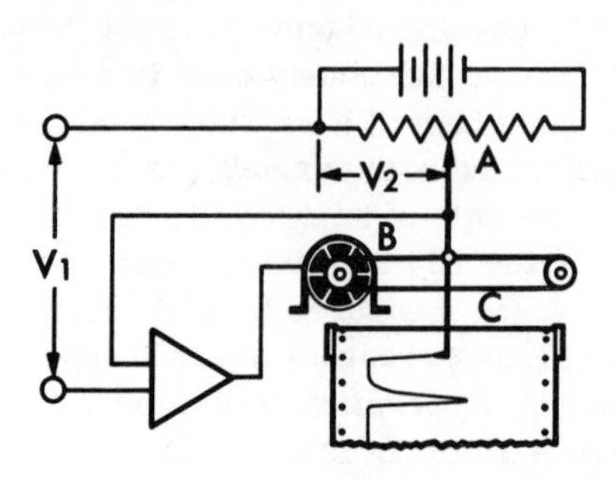

FIGURE II-4—SCHEMATIC OF POTENTIOMETRIC RECORDER

Recorder characteristics which affect the ability of the recorder to faithfully plot the input signal, i.e., can affect quantitative accuracy, are discussed below.

Dead-band. The range through which the measured quantity can be varied without causing a readable graphic response is defined as dead-band, usually expressed in percent of full-scale. Mechanical load and amplifier gain are the two factors which cause dead-band.

Range. The region covered by the two end-scale values, is the range of the recorder, 1 mv. is the standard G.C. recorder range.

Zero shifting. Shifting of the zero point of the recorder may be observed with recorders which have inadequate shielding from the a.c. circuits.

Pen speed. The time required to record a full-scale step change of signal applied to the recorder is referred to as the pen speed. Most chromatographic recorders have a full-scale response of 1 second or less.

Linear range. The recorder's linear range is equal to the ratio of a 100% peak to that of the smallest discernible peak, typically about 0.5%. Of course, the linear range is effectively extended by the attenuation of the input signal.

BIBLIOGRAPHY

1. *Giddings, J.C., Journal of Chemical Education, 39, No. 11, 569-573 (1962).*

III. THEORY

A. INTRODUCTION

The resolution of chromatographic peaks is related to two factors: column efficiency, and solvent efficiency. The former is concerned with the peak broadening of an initially compact band as it passes through the column. The broadening results from the column design and operating conditions, and can be quantitatively described by the height equivalent to a theoretical plate (HETP). The HETP is that length of column necessary for the attainment of solute equilibrium between the moving gas phase and the stationary liquid phase.

Solvent efficiency on the other hand results from the solute-solvent interaction and determines the relative position of solute bands on a chromatogram. Solvent efficiency, or relative retention, is expressed as the ratio of peak maxima (adjusted retention times). This is determined by the respective distribution coeficients of the solutes in the solvent at a given temperature. Figure III-1 illustrates both increased column efficiency and solvent efficiency.

B. COLUMN EFFICIENCY

Column efficiency is measured by the number of theoretical plates. To compare column efficiencies, one must specify the solvent, solute, temperature, flow rate, and sample size.

The plate concept is a carryover from distillation processes where historically, the first efficient columns were in fact composed of discrete plates. In

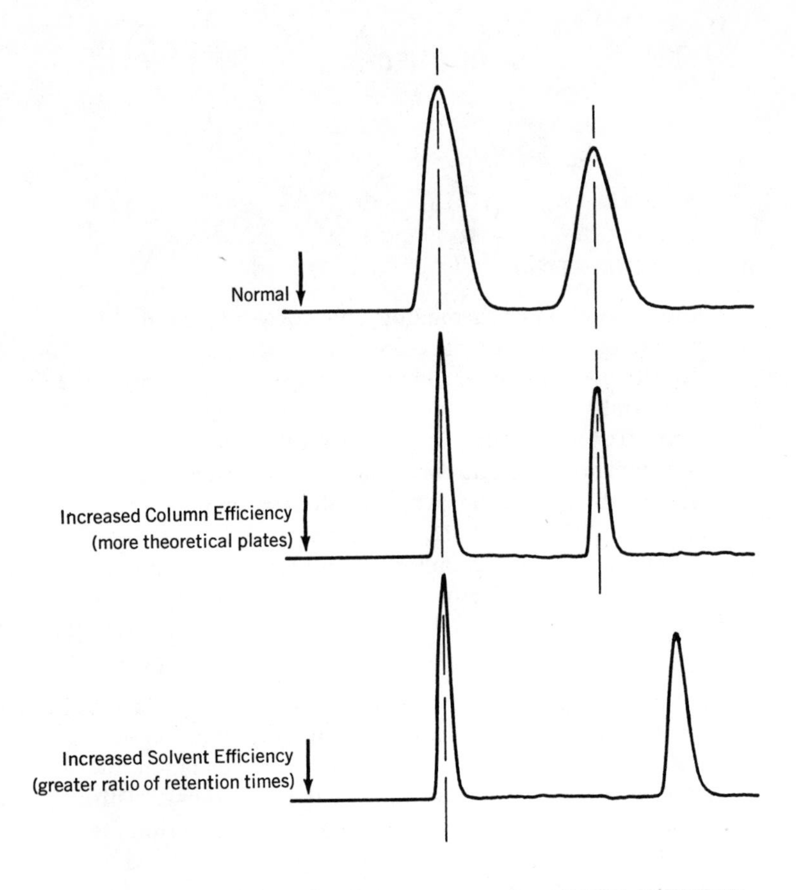

FIGURE III-1—ILLUSTRATION OF COLUMN AND SOLVENT EFFICIENCY

the laboratory, however, packed distillation columns are more commonly employed and the plate value is a theoretical concept introduced so that column performance can be evaluated.

In Gas Chromatography, as in distillation the discrete plate is an artificial concept. The number of plates obtained in the methods varies greatly and should not be taken as a direct measure of the difference in their separating power. Plates are useful to compare similar columns, or set standards for packing techniques.

Theoretical plates can be easily measured from the chromatogram. Tangents are drawn to the peak at the points of inflection (about 2/3 of the height). The number of theoretical plates, N, is given by 16 $(x/y)^2$, where "y" is the length of the baseline cut by the two tangents, and "x" is the distance from injection to peak maximum (including the dead-volume).

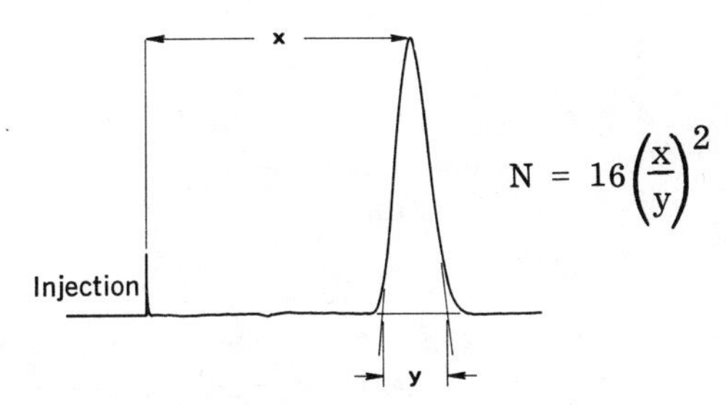

FIGURE III-2—CALCULATION OF THEORETICAL PLATES

Many factors affect column efficiency and most of these have been evaluated by their effect on N, or the height equivalent to a theoretical plate, HETP. This is related to N by:

$$\text{HETP} = L/N$$

where L is the length of the chromatographic column, usually in centimeters. HETP calculation allows comparisons between columns of different lengths and is the preferred measure of column efficiency.

1. <u>Rate Theory</u>

   Several chromatographic theories have been developed to account for the shape of elution curves from chromatographic columns. These will not be discussed here. It is recommended that serious students examine a rate theory developed by van Deemter et al [1]. This theory was extended by Glueckauf [2] and other workers. A qualitative understanding of the van Deemter equation is useful in optimizing chromatographic performance.

The three principal contributions to the broadening of a band are:

a. Multipath effect or eddy diffusion (A term).
b. Molecular diffusion (B term).
c. Resistance to mass transfer (gas and liquid, C term).

From these a basic equation can be derived [1,3] for the height equivalent to a theoretical plate in a gas-liquid column:

$$\text{HETP} = A + B/\mu + c \bullet \mu$$

where A, B, and C are the above mentioned constants and μ the linear gas velocity (or flow rate) through the chromatographic column. The linear gas velocity is measured by:

$$\mu = \frac{\text{length of column, cm}}{\text{retention time of air, seconds}}$$

If HETP is plotted against μ one obtains a hyperbola with a minimum HETP. This minimum is that flow rate (μ optimum) at which the column is operating most efficiently. However, owing to the compressibility of the carrier gas, μ is not constant over the entire length of the column, hence only a small section can operate at maximum efficiency. The influence of the parameters of the equation on the efficiency of separation has been discussed by Keulemans [3,4] and others [1,5,6].

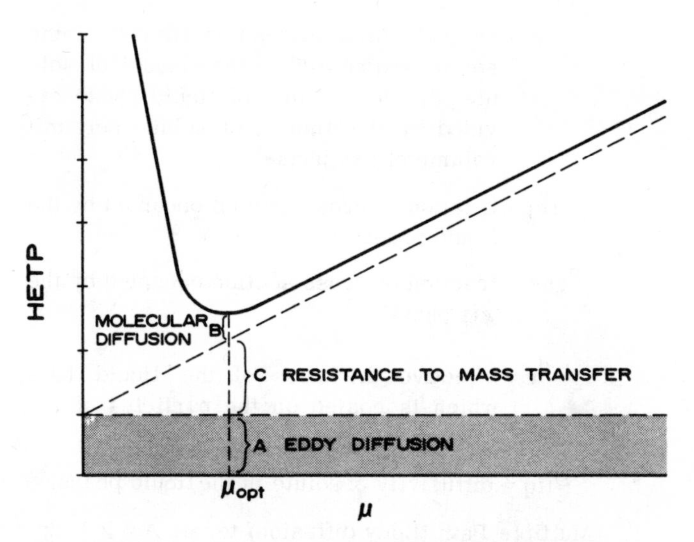

FIGURE III-3—PLOT OF HETP AGAINST GAS VELOCITY

The expanded version of the Van Deemter equation is:

$$\text{HETP} = 2\lambda\, dp + \frac{2\,\gamma D_{gas}}{\mu} + \frac{8}{\pi^2}\,\frac{k'}{(1+k')^2}\,\frac{d_f^2}{D_{liq}}\;\mu$$

where:

λ = a constant which is a measure of packing irregularities.

γ = a correction factor accounting for the tortuosity of the gas channels in the column.

dp = average particle diameter of the solid support.

D_{gas} = diffusivity of the solute in the gas phase

μ = the linear gas velocity

k' = capacity factor = $k(F_{liq}/F_{gas})$

k = partition (distribution) coefficient of the solute, expressed as the amount of solute per unit volume of liquid phase divided by the amount of solute per unit volume of gas phase

F_{liq} = fraction of cross section occupied by the liquid phase

F_{gas} = fraction of cross section occupied by the gas phase

d_f = effective thickness of the liquid film which is coated on the particles of the support

D_{liq} = diffusivity of solute in the liquid phase.

2. Multiple Path (Eddy diffusion) term, $A = 2\ \lambda\ dp$
In any packed column solute molecules and carrier gas molecules travel along many paths. These paths have different lengths, therefore the solute molecules have different residence time. This adds to peak broadening. This broadening depends upon the size of the particles constituting the packing, the shape, and the manner in which they are packed and the column diameter.

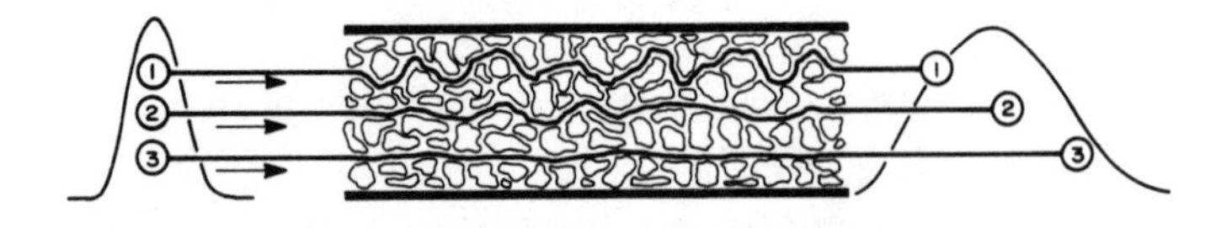

FIGURE III-4—ILLUSTRATION OF MULTIPATHS

One obvious way of decreasing the A term would seem to be in using smaller particles. The λ constant, however, characterizes the manner in which the particles are packed. According to Klinkenberg and Sjenitzer[8] it is easier to obtain regular packing with large rather than small particles.

TABLE III-1—EFFECT OF MESH SIZE ON λ

MESH SIZE	λ
200-400 (dp = 0.07-0.04 mm)	~ 8
50-100 (dp = 0.3-0.15 mm)	~ 3
20-40 (dp = 0.8-0.4 mm)	~ 1

The high value for λ (Table III-1) in the small mesh sizes may be due to channeling.

Another limiting factor on particle size is the pressure drop across the column. Small particles increase the pressure drop. Since the linear gas velocity will be more uniform at low rather than high inlet-to-outlet pressure ratios, it is desirable to operate the whole column at the linear gas velocity corresponding to optimum efficiency.

To minimize the A term (increase column efficiency) one should use small particles of uniform size consistent with low pressure drop and low λ and smaller diameter columns. The packing technique should give a high packing density without crushing the particles.

3. Molecular Diffusion term, $B = 2\,\gamma D_{gas}/\mu$
The molecular diffusion term is proportional to D_{gas}, the solute diffusivity in the carrier gas. High solute diffusivity leads to band broadening

and a consequent loss in the efficiency. It is a property of both solute and carrier gas and may be reduced by increasing the density of the gas, either by increasing the pressure or molecular weight of the gas. The use of a dense carrier gas such as nitrogen or argon would be preferred over hydrogen and helium. There may be other reasons, such as detector requirements, or analysis time, why a light carrier gas is preferred.

Solute diffusion in the liquid phase is extremely small compared to that in gas phase and can be neglected.

The tortuosity correction, γ, adjusts the linear velocity to the larger actual velocity in a packed column. It is the difference between a straight line path and the average real path of a molecule. With increasing particle size, γ increases to a limiting value of 1.

The method of reducing the B term is to increase the linear flow velocity μ, and use a high molecular weight carrier gas.

4. <u>Resistance to mass transfer term,</u>

$$C = \frac{8}{\pi^2} \frac{k'}{(1 + k')^2} \frac{d_f^2}{D_{liq}} \bullet \mu$$

The important effect of the amount of liquid on the solid support shows up in the C term, in which the thickness of the film d_f enters in the second power. Obviously keeping d_f small will reduce the C term.

d_f also affects the quantity k' where $k' = K\, F_{liq}/F_{gas}$. This complicates the influence of d_f on C. In any given column $F_{liq} + F_{gas}$ = constant, so that a change in d_f changes F_{liq}, F_{gas}, and F_{liq}/F_{gas}.

Thus the C term is a complex function of the film thickness.

A liquid phase which exhibits high diffusivity, D_{liq}, tends to reduce this C term. For this reason liquids of low viscosity produce more efficient columns.

To minimize the C term, a thin uniform film of a low viscosity liquid should be used. The flow rate must be low enough and distribution coefficient high enough to favor equilibrium between the liquid and gas phase.

5. Summary

The conclusions which have been drawn from the Rate Theory of Van Deemter are of practical interest and can be used to improve column efficiency:

a. Particle diameter: Column efficiency is improved by the use of small, uniform particle sizes. Supports which give the highest efficiency are diatomaceous earth type with 100-120 mesh range.

b. Flow rate: For maximum efficiency the column must be operated at the optimum flow rate. This is found on a plot of HETP against the flow rate. The minimum HETP determines the optimum linear gas velocity or flow rate to be used. In practice, operating at flow rates slightly higher than optimum will decrease the analysis time and not materially affect the HETP.

c. Carrier gas: The detector employed usually dictates the choice of carrier gas. However, for highest efficiency a high molecular weight gas should be the choice. Where rapid analysis time is required and highest efficiency is not necessary a low molecular weight carrier gas such as helium or hydrogen would be preferred.

d. Type of liquid phase: A low viscosity, low vapor pressure solvent with good absolute solubility for the sample should be used. To obtain a separation, it must also exhibit a differential solubility.

e. Amount of liquid phase: Low liquid loadings (thin film) 1% - 10% have the advantage of fast analysis and lower temperature operation. Low liquid loadings however reduce the sample capacity and may require highly inactive solid supports.

f. Pressure: The majority of practical gas chromatographers work at an outlet pressure of one atmosphere so that operation at the optimum flow rate fixes the inlet pressure. However, best efficiency is obtained at low inlet-to-outlet pressure ratios.

g. Temperature: Resolution can usually be improved by lowering the column temperature. The practical limit to low temperature is the resulting long analysis time. Lowering the temperatures also decreases decomposition of the compounds but may increase adsorption. Simultaneous reduction of liquid loading and temperature is generally beneficial. The limit to this is the increased exposure of active sites on the solid support and the need for sensitive detectors.

h. Column diameter: Capillary and preparative column experiments indicate that efficiency is improved with decreasing internal diameter. Thus 1/8 inch OD and even 1/16 inch

OD columns are used for highest resolution over the formerly more common 1/4 inch OD columns.

C. SOLVENT EFFICIENCY

A striking feature of gas chromatography is that substances having the same vapor pressure can be easily separated by appropriate selection of the liquid phase. The major advantage of G.C. over distillation is the ability to use selective solvents. There are numerous selective phases available and the following comments should aid in selecting the proper one.

1. <u>Interaction forces and partition coefficient</u>
   There are four interaction forces which can aid in the G. C. separation[4,8]:

   a. Orientation or Keesom forces: Forces resulting from the interaction between two permanent dipoles. The "hydrogen-bond" is a particularly important type of orientation force encountered in gas chromatography.

   b. Induced dipole, or Debye forces: Forces resulting from the interaction between a permanent dipole in one molecule and the induced dipole in a neighboring molecule. These forces are usually very small.

   c. Dispersion, London or non-polar forces: Forces arising from synchronized variations in the instantaneous dipoles of the two interaction species. These forces are present in all cases, and are the only source of attraction energy between two non-polar substances. They are weak compared to "a" and "b".

   d. Specific interaction forces: Forces resulting from chemical bonding, complex formation between solute and solvents.

These forces of interaction determine the separation achieved. Their combined effects are expressed by the partition coefficient k, where:

$$k = \frac{\text{amount of solute per unit volume of liquid phase}}{\text{amount of solute per unit volume of gas phase}}$$

The value of k is high when most of a substance is retained in the liquid phase. This means that the substance moves slowly down the column because only a small fraction will be in the carrier gas at any given time. Transport is negligible in the liquid phase, and only that fraction in the gas phase is carried through the column.

Thus, separation between two compounds is possible, if their partition coefficients are dissimilar. The greater the difference in their k values, the fewer the plates or the shorter the column length that is required to achieve a separation.

2. Solvent efficiency and temperature
Solvent efficiency is measured by α, the relative retention. It is the ratio of adjusted retention

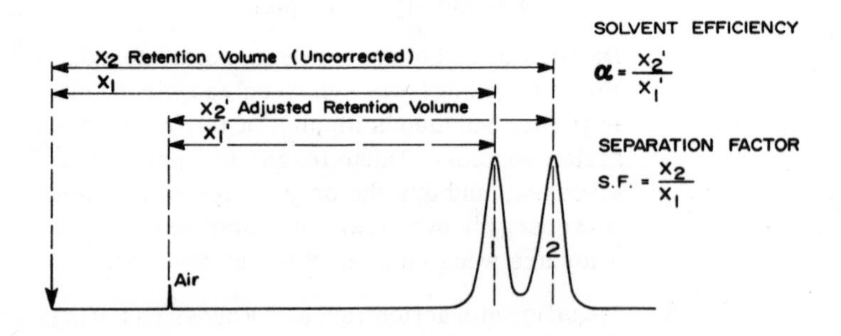

FIGURE III-5—CALCULATION OF SOLVENT EFFICIENCY

times or partition coefficients. Relative retention differs from the separation factor S.F., where S. F. is the ratio of uncorrected retention times.

Both α and k are temperature dependent. However, over a limited temperature range α will be constant. The distribution coefficient, k, decreases with increasing temperature, i.e., the fraction of the solute in the gas phase will increase and hence the elution time will decrease. This results in decreased separation since it is the liquid phase which performs the separation. No separation occurs in the gas phase. To achieve better separations, lower temperatures should be used. Lower temperatures mean more liquid phase interaction, more separation, longer analysis time. As a minimum, the solute should spend 50% of the time in the liquid phase, so that the retention time exceeds twice the retention time of air.

3. Resolution

The true separation of two consecutive peaks is measured by the resolution, R. Resolution is a measure of both the column and solvent efficiencies. It accounts for both the narrowness of peaks and the separation between maxima.

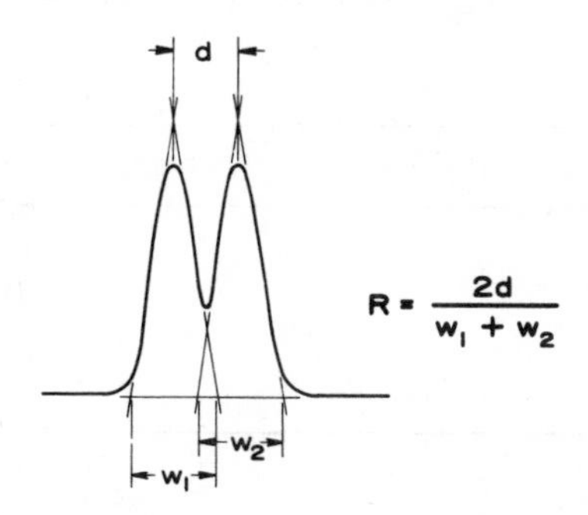

FIGURE III-6—CALCULATION OF RESOLUTION

If $R = 1$, the resolution of two equal-area peaks is approximately 98% complete.

If $R = 1.5$, baseline separation (99.7% resolution) is achieved.

4. Number of plates for required separation
   An equation which is useful in determining the number of plates and thus the length of column required is:

$$N_{req} = 16\,R^2 \left(\frac{\alpha}{\alpha - 1}\right)^2 \left(\frac{k'_2 + 1}{k'_2}\right)^2$$

where k'_2 = capacity factor for peak 2

$$= \frac{\text{adjusted retention time, } X'_2}{\text{retention time air}}$$

R is the resolution required, and α is the solvent efficiency, both as defined above.

Example: The separation shown below was obtained with a 3 meter column. What is the minimum column length necessary to obtain a resolution of 1.5?

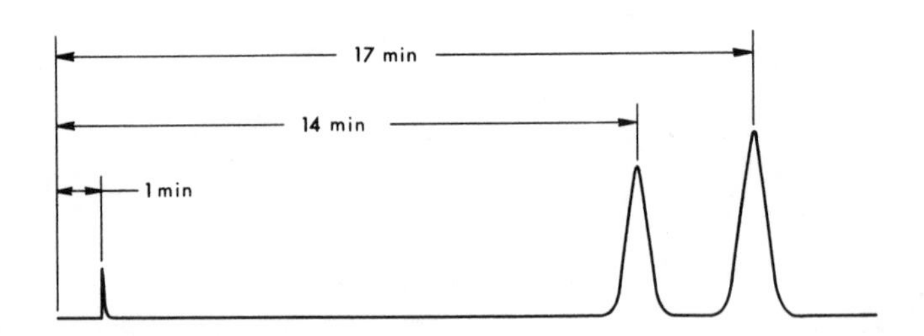

FIGURE III-7—CALCULATION OF REQUIRED PLATES

$$N_{orig} = 16\left(\frac{x}{y}\right)^2 = 16\left(\frac{17}{1}\right)^2 = 3024 \text{ plates}$$

$$\alpha = \frac{17-1}{14-1} = 1.231; \quad k_2 = \frac{17-1}{1} = 16$$

$$N_{req} = 16\, R^2 \left(\frac{\alpha}{\alpha - 1}\right)^2 \left(\frac{k_2' + 1}{k_2'}\right)^2$$

$$= 16(1.5)^2 \left(\frac{1.231}{1.231-1}\right)^2 \left(\frac{16 + 1}{16}\right)^2$$

$$= 36\ (28.4)\ (1.13) = 1155 \text{ plates}$$

$$L_{reqd} = L_{orig} \times \frac{N_{req}}{N_{orig}}$$

$$= 3 \times \frac{1155}{3024} = 1.14 \text{ meters}$$

Thus a much shorter column will provide a satisfactory separation. Since the separation is complete on the 3 meter column, higher flow rates can be used to reduce the analysis time.

BIBLIOGRAPHY

1. *Van Deemter, J.J., Zuiderweg, F.J., and Klinkenberg, A., Chem. Eng. Sci., 5, 271 (1956).*
2. *Gluechauf, E., Ion Exchange and Its Applications, Soc. of Chem. Ind., London, p. 34, 1955.*
3. *Keulemans, A.I.M., and Kwantes, A., V.P.C. Symp., Butterworths, London, p. 15, 1957.*
4. *Keulemans, A.I.M., Gas Chromatography, Reinhold Publ. Corp., New York, 1957.*
5. *Van Deemter, J.J., Gas Chrom. Discuss. Group, Cambridge, Engl., Oct. 4, 1957.*
6. *Littlewood, A.B., Preprint T208, 2nd Gas Chrom. Symp., Amsterdam, May 19, 1958.*
7. *Keulemans, A.I.M., Gas Chromatography, Reinhold Publ. Corp., New York, 1957.*
8. *Klinkenberg, A., and Sjenitzer, F., Chem. Eng. Sci., 5, 258 (1956).*
9. *Keulemans, A.I.M., Kwantes, A., and Zaal, P., Anal. Chim. Acta., 13, 357 (1955).*

IV. COLUMNS, THEORY AND TECHNIQUE

A. INTRODUCTION

The column is the heart of the chromatograph. The actual separation of sample components is achieved in the column. Consequently, the success or failure of a particular separation will depend to a large extent on the choice of column. In gas-liquid chromatography, there are both capillary and packed columns. Capillary columns are open tubes of small diameter with a thin liquid film on the wall. Packed columns consist of an inert solid material supporting a thin film of a nonvolatile liquid. The tube may be either glass, metal or plastic, coiled to fit the chromatograph oven. The solid support, type and amount of liquid phase, method of packing, length, and temperature of the column are important factors in obtaining the desired resolution. The dimensions of the column govern the total amount of gas and liquid it will contain. Analytical columns are ordinarily 1/16" to 1/4" O.D. tubing from 3 to 30 feet in length. Lengthening a column will increase the separation, sometimes substantially.

B. LIQUID PHASE

There is no fool-proof method for selecting the best liquid phase for a particular separation. The right selection is based mainly on experience and/or trial and error. Considerable column technology has been reported in the literature, so a short survey could save considerable experimental time.

1. Liquid phase requirements

a. Good absolute solvent for sample components - if solubility is low, components elute rapidly, and separation is poor.
b. Good differential solvent for sample components.
c. Nonvolatile - vapor pressure of 0.01 to 0.1 mm at operating temperature for reasonable column life.
d. Thermally stable - instability can be promoted by catalytic influence of the support as temperature increases.
e. Chemically inert toward the solutes of interest at the column temperature.

2. <u>Choice of liquid phase</u>

The liquid phase chosen depends on the composition of the sample. Hopefully, the type components likely to be present in the sample will be known before the analysis is started. The more one knows about a sample (suspected components, boiling range, structures) the easier it is to select the proper column and operating conditions.

For an efficient, normal separation, the liquid phase should be similar in chemical structure to the components of the mixture.

Example: Hydrocarbon compounds are best separated with a hydrocarbon solvent: Paraffins on squalane (a long chain hydrocarbon); polar compounds with a polar solvent: alcohols on Hallcomid (an amide).

If the components of the mixture are of different chemical classes, but close in boiling point, liquid phases of different polarity must be used. By varying the polarity of the solvent, interaction forces may be brought into play to effect a separation. These forces are described in Chapter III and should be reviewed if they are unfamiliar to the student.

Table IV-1 shows the various functional groups in the five classifications suggested by Ewell et al.[6]

TABLE IV-1—SOLUTE CLASSIFICATION

CLASS I	CLASS II
Water	Alcohols
Glycol, glycerol, etc.	Fatty acids
Amino alcohols	Phenols
Hydroxy acids	Primary and secondary amines
Polyphenols	Oximes
Dibasic acids	Nitro compounds with α-H atoms
	Nitriles with α-H atoms
	NH_3, HF, N_2H_4, HCN
CLASS III	**CLASS IV**
Ethers	$CHCl_3$
Ketones	CH_2Cl_2
Aldehydes	CH_3CHCl_2
Esters	CH_2ClCH_2Cl
Tertiary amines	$CH_2ClCHCl_2$ etc.
Nitro compounds with no α-H atoms	Aromatic hydrocarbons
Nitriles with no α-H atoms	Olefinic hydrocarbons
CLASS V	
Saturated hydrocarbons	
CS_2	
Mercaptans	
Sulfides	
Halocarbons not in Class IV such as CCl_4	

Class I consists of compounds capable of forming networks of hydrogen bonds. Class II is composed of compounds containing both a donor atom

(O, N, F) and an active hydrogen atom. Molecules containing donor but no active hydrogen atoms are in Class III. Class IV is made up of molecules containing an active hydrogen but no donor atoms. Compounds exhibiting no hydrogen bonding capacity are placed in Class V.

Table IV-2 shows some commonly used liquid phases arranged in the classification given in Table IV-1.

TABLE IV-2—LIQUID PHASE CLASSIFICATION

CLASS A (I)	CLASS B (II)
FFAP	Tetracyanoethyl pentaerythritol
20M-TPA	Zonyl E-7
Carbowaxes	Ethofat
Ucons	β, β-Oxydipropionitrile
Versamid 900	XE-60
Hallcomid	XF-1150
Quadrol	Amine 220
Theed	Epon 1001
Mannitol	Cyanoethyl sucrose
Diglycerol	
Castorwax	

CLASS C (III)	CLASS D (IV & V)
All polyesters	SE-30
Dibutyl tetrachlorophthalate	SF-96
SAIB	DC-200
Tricresyl phosphate	Dow 11
STAP	Squalane
Benzyl cyanide	Hexadecane
Lexan	Apiezons
Propylene carbonate	
QF-1	
Polyphenylether	
Dimethylsulfolane	

A solute will be retained more strongly by that liquid phase, which is closest according to its classification in Table IV-2. This means higher solubility and usually better separation.

TABLE IV-3—EFFECT OF CLASSES ON RETENTION

Solute - Liquid	
III - C IV - D V - C or D	Quasi-ideal systems; solutes separate according to boiling points.
IV - C	Solute well retained by liquid phase.
I, II, III > A or B	Solute generally retained by liquid phase.
II - D IV, V > B	Solute not well retained by liquid phase.
I - D IV, V > A	Solute not retained by liquid phase, frequently very limited solubility.

Consider ethanol (b.p. 78°C, Class II) and 2,2-dimethylpentane (b.p. 79°C, Class V). For good separation, choose a liquid phase from B (or A) or one at the other end of the scale, D. With a solvent from B, the ethanol will be selectively retained and the paraffin will emerge first. With a solvent from D, the order will be reversed, and the alcohol will emerge first.

If, in addition, a third material--benzene (b. p. 80°C, Class IV) were present, it would be preferable to use a polar solvent from B or A. Then a small dipole would be induced in the slightly polarizable benzene, retaining it a little longer than the paraffin.

Elution order would be paraffin, benzene, and ethanol. Not only orientation forces, but induction forces are responsible for this separation.

3. "Super-selective" liquid phases

Certain liquid phases may be said to be "super-selective". Some of these liquids depend on the formation of loose chemical adducts with particular solutes to effect resolution.

a. Silver nitrate complexes

Silver ions form loose adducts with olefins of the type shown below.

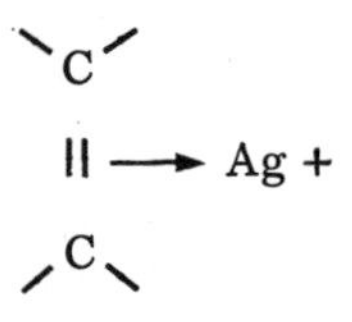

Saturated solutions of silver nitrate in ethylene glycol, glycerol, and triethylene glycol (1, 8) or in benzyl cyanide (11) have been used as selective solvents for olefins. Benzyl cyanide has an advantage over the glycols, because it is nonhygroscopic and does not require moisture-free carrier gas. Paraffins are not retained and pass rapidly through the column, while alkynes react to form silver acetylides and remain on the column. Temperatures should be below 40°C; above this limit the adducts with $AgNo_3$ are not formed, and solutions are unstable.

b. Mercuric perchlorate complexes

Mercuric perchlorate has been used to remove unsaturates by π -complexes with olefins, alkynes, and aromatics [7].

c. Tetracyanoethylated pentaerythritol (TCEPE)

TCEPE is highly selective for aromatics. Paraffins to C_{13}, α -olefins to C_{13}, cyclo-olefins to C_{10}, cycloparaffins to C_{10} are eluted as a group before benzene.

d. Bentone-34 for o, m, p-Isomers [2,10,12,13]

Bentone-34 is a montmorillonite clay in which the naturally occurring cations have been exchanged for dimethyldioctadecyl ammonium ions. The clay is mixed with a phthalate ester or silicone oil. This mixture is used to coat the solid support. The three xylene isomers and ethylbenzene are resolved on a 5% Bentone-34, 5% Dinonylphthalate on 80/100 Chromosorb W column, 15 feet x 1/8 inch at 110°C. The diethylbenzene, ethyltoluene, and dichlorobenzene isomers have been separated on Bentone-34 columns. The selective nature of Bentone-34 can be attributed to its layer structure. The ease of adsorption of the solute between the layers of the expanded clay presumably depends on the geometrical shape of the solute molecules. Bentone-34 does not retain para isomers, so they are eluted ahead of m- and o-isomers.

e. Liquid crystals

Liquid crystals are intermediate between the crystalline solids and "normal" isotropic liquids. They can exist in liquid crystal phases --smectic and nematic. The solid passes from the smectic liquid phase to the nematic, and finally to the "normal" liquid phase as temperature is raised [4,5]

Molecules in the nematic phase are free to move about only so long as they remain parallel to one another. This organized structure allows a nematic liquid crystal to show a selective affinity for linear molecules. They will retain p-disubstituted benezenes relative to o - and m- isomers.

A typical formula for a liquid crystal is:

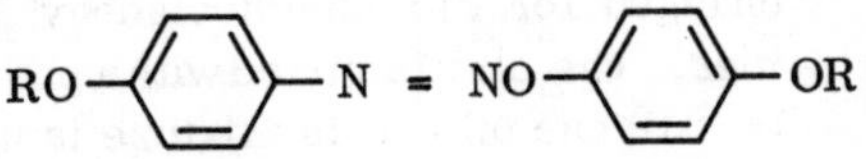

RO—⟨benzene ring⟩—N = NO—⟨benzene ring⟩—OR

where R is CH_2 or n-C_6H_{13} or n-C_7H_{15}

Para-azoxyanisole is a nematic liquid between 120° and 135^oC.

TABLE IV-4—RECOMMENDED LIQUID PHASES BY SAMPLE TYPE

CLASSIFICATION OF COMPOUNDS	STATIONARY PHASE
ACIDS	
C_1-C_{18} (free)	FFAP
Bile and Urinary	SE-30
Fatty Acid-Methyl Esters	DEGS
	FFAP
	Apiezon L
	TCEPE
	EGSS -X

TABLE IV-4—RECOMMENDED LIQUID PHASES BY SAMPLE TYPE (cont.)

CLASSIFICATION OF COMPOUNDS	STATIONARY PHASE
ALCOHOLS	
C_1-C_5	Hallcomid M-18 OL
	Carbowax 600 or 1540
C_1-C_{18}	FFAP
	Carbowax 20M
Di-Poly	FFAP
	QF-1
ALDEHYDES	
C_1-C_5	Ethofat
C_5-C_{18}	Carbowax 20M
ALKALOIDS	QF-1
	SE-30
AMINO ACID DERIVATIVES	
$\overline{N}$- Butyl trifluoroacetyl esters	DEGS/EGSS-X
AMINES	
See Nitrogen Compounds	
BORANES	Apiezon L
ESSENTIAL OILS	
General	FFAP
	Carbowax 20M
ESTERS	
Mixed	Dinonyl phthalate
	Porapak Q
ETHERS	Carbowax 20M
GLYCOLS	Porapak Q
HALOGEN COMPOUNDS	Carbowax 20M
	QF-1 (FS-1265)
	FFAP
Freons	Dibutyl Tetrachloro-phthalate
	UCON Polar 2000
HYDROCARBONS	
Aliphatic	
C_1-C_5	Propylene Carbonate
	Carbowax 400
	Tributyl phosphate
C_5-C_{10}	Didecylphthalate
	SE-30
Aromatics	Tetracyanoethylated Pentaerythritol
	Dibutyl tetrachloro-phthalate

TABLE IV-4—RECOMMENDED LIQUID PHASES BY SAMPLE TYPE (cont.)

CLASSIFICATION OF COMPOUNDS	STATIONARY PHASE
Hydroxy	2,4 Xylenyl Phosphate
Olefins	
C_1-C_6	$AgNO_3$/Benzyl cyanide
	Dimethylsulfolane
	Propylene Carbonate
C_6-up	Carbowax 20M
Polynuclear	SE-30 on DMCS treated support
	FFAP on DMCS treated support
	PMPE (5 ring)
KETONES	Lexan
	FFAP
NITROGEN COMPOUNDS	
Amines	Dowfax 9N9/KOH
Amides	Versamid 900
Ammonia	Ethofat or Carbowax 600 on Chromosorb T
NITRILES	Tetracyanoethylated pentaerythritol
	FFAP
	XF-1150
ORGANO METALLIC	FFAP
	SE-30
PESTICIDES	Dow 11
	QF-1 (FS-1265)
	SE-30
	OV-1 OV-17
PHOSPHORUS	SE-30
	STAP
SILANES	SF-96
	FFAP
STEROIDS	STAP
	XE-60
	QF-1 (FS-1265)
	SE-30
	OV-1 OV-17
SUGAR DERIVATIVES	
Trimethylsilylethers	QF-1
	SE-52
SULFUR COMPOUNDS	Carbowax 20M
	FFAP
	Dinonylphthalate
	Porapak Q

TABLE IV-4—RECOMMENDED LIQUID PHASES BY SAMPLE TYPE (cont.)

CLASSIFICATION OF COMPOUNDS	STATIONARY PHASE
WATER	Porapak Q
GASES	
Ar and O_2	6-feet Activated Molecular Sieve 5A at -72°C
H_2, O_2, He, N_2, CO, CH_4	20-feet Molecular Sieve 13X
CO_2, H_2S, CS_2, COS	Silica Gel or Porapak Q
H_2 Isotopes	6-feet Activated Molecular Sieve 13X
CO_2, N_2, O_2, CH_4, CO	2-1/2 feet Silica Gel internal column 20-feet Molecular Sieve 13X external column
N_2O, CO_2, NO	Porapak Q

5. Special cases

The following chromatograms illustrate the difficulty sometimes encountered in the selection of the right liquid phase for a particular separation. The liquid phase listed beside each trace was the only one found that was capable of making the separation.

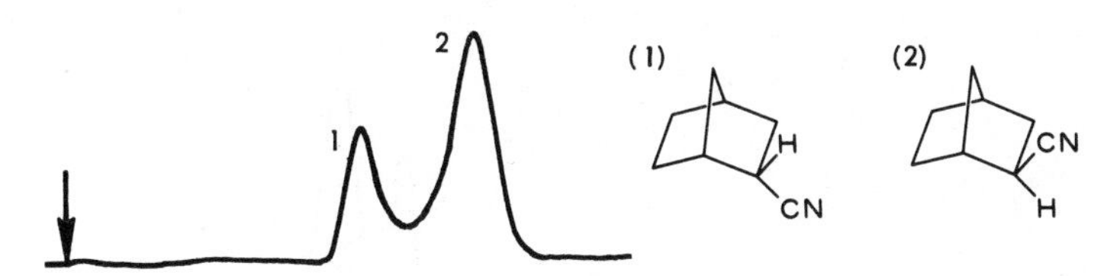

10 feet 20% FFAP on 60/70 mesh DMCS-AW Chromosorb W at 140°C

FIGURE IV-1a—EXO & ENDO NORBORNYLS

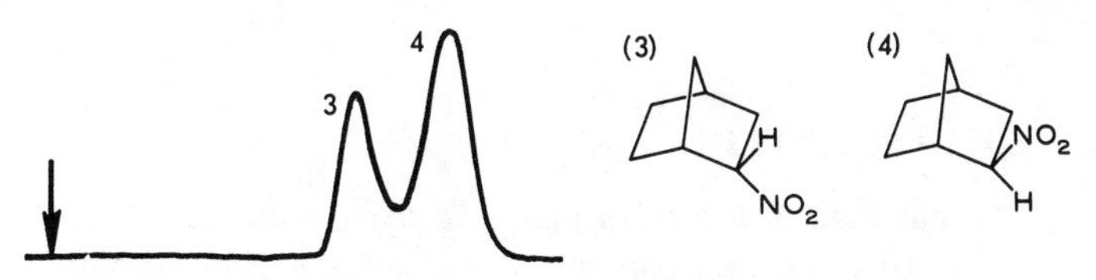

10 feet 20% FFAP on 60/70 mesh DMCS-AW Chromosorb W at 140°C

FIGURE IV-1b—EXO & ENDO NORBORNYLS

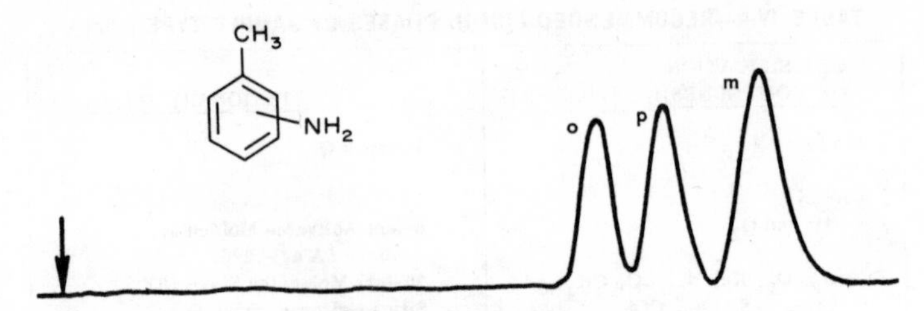

10 feet 5% Carbowax 20M on 60/80 mesh KOH-treated Chromosorb W at 140°C

FIGURE IV-1c—o,m,p-TOLUIDINES

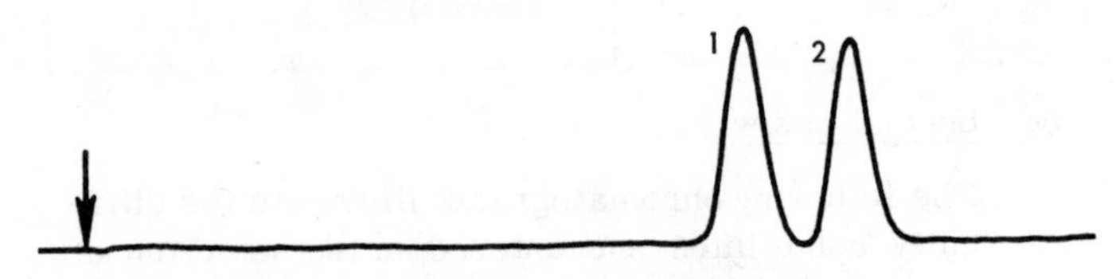

20 feet 30% TCEPE on 60/80 mesh DMCS-AW-W at 105°C

FIGURE IV-1d—1. 2,5 DIMETHYL, 2,5-(DI-TERT-BUTYL PEROXIDE) HEXYNE
2. 2,5 DIMETHYL, 2,5-(DI-TERT-BUTYL PEROXIDE) HEXANE

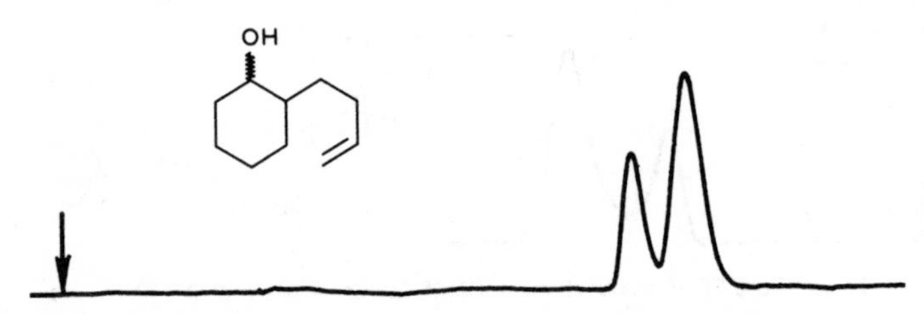

10 feet 15% Zonyl E-7 on 60/80 AW-Chromosorb W at 120°C

FIGURE IV-1e—EPIMERS

C. SOLID SUPPORTS

The purpose of the solid phase is to support a thin uniform film of liquid phase. An optimum support should have certain characteristics:

1. A large specific surface area--from 1 to 20 sq. meters ^{2}/gram
2. A pore structure with uniform pore diameter in the range of 10μ or less
3. Inertness--a minimum of chemical and adsorptive interaction with the sample
4. Regularly-shaped particles, uniform in size for efficient packing
5. Mechnical strength--should not crush on handling

No material has yet been described which fills all these requirements; however, several suitable supports are commercially available. Usually one must choose between inertness or efficiency (high surface area).

The raw material for most gas chromatographic supports is diatomite; also known as diatomaceous silica, diatomaceous earth, and the German word - Kieselguhr. Diatomite is composed of the skeletons of diatoms, microscopic unicellular algae, which are primarily micro-amorphous hydrous silica, although some minor impurities, mainly metallic oxides are present. The individual diatom skeletons are very small, highly porous and have a high surface area.

There are five forms or types of Chromosorb®*,A, G, P, W and T. Each is available either untreated or treated and in a variety of mesh ranges.

Chromosorb A is for use in preparative-scale gas chromatography. It has good capacity to hold liquid phase (25% maximum), a structure that does not readily break down with handling and a surface that is not highly adsorptive. It is available in mesh ranges of 10/20, 20/30 and 30/40. These allow the use of long columns with a low-pressure drop.

Chromosorb G for the separation of polar compounds. Its low surface area, hardness and good handling characteristics make it a good replacement for light, friable supports such as Chromosorb W. Because of its

**Chromosorb is Johns Manville's registered trademark for G.C. support material*

lower surface area and higher density, Chromosorb G is employed with a lower liquid phase coating. A 5% coating on Chromosorb G corresponds to 12% on Chromosorb W.

Chromosorb P is prepared from the production of Johns - Manville's Sil-O-Cel C-22 Firebrick. It is a calcined diatomite, pink in color and relatively hard. Its surface is more adsorptive than the other Chromosorb grades and is used primarily for hydrocarbon work. It has the best efficiencies for hydrocarbons.

Chromosorb T is a fluorcarbon resin screened from DuPont Teflon 6. It is recommended for use in the separation of highly polar or reactive compounds such as water, hydrazine, sulfur dioxide and halogens. Its surface is inert to these compounds and symmetrical peaks are obtained. Because of its relatively poor column efficiency, Chromosorb T is recommended only when its highly inert surface is required.

Chromosorb W is a flux-calcined diatomite support prepared from the production of Johns - Manville Celite filter aids such as Celite 545. It is similar in performance and properties to Celite 545. Chromosorb W is white in color and friable compared to Chromosorb G. Its surface is relatively nonadsorptive and is used for the separation of polar compounds.

TABLE IV-5—CHEMICAL ANALYSIS OF SUPPORTS[3]

	Firebrick C22	Celite 545	Chromosorb P	Chromosorb W
SiO_2	89.7	89.9	89.2	91.2
Al_2O_3	5.1	3.6	5.1	4.1
Fe_2O_3	1.55	1.65	1.50	1.15
TiO_2	0.30	0.30	0.30	0.25
CaO	1.30	1.75	0.90	0.40
MgO	0.90	0.70	1.00	0.65

TABLE IV-6—PHYSICAL PROPERTIES OF SUPPORTS

Type	Free Fall Density	Packed Density	Surface Area m^2/g	pH
Chromosorb T	0.42	0.49	7.8	--
Chromosorb G	0.47	0.58	0.5	8.5
Chromosorb W	0.18	0.24	1.0	8.5
Chromosorb A	0.40	0.48	2.7	7.1
Chromosorb P	0.38	0.47	4.0	6.5

The diatomite support surfaces are covered with silanol (Si-OH) and siloxane groups (Si-O-Si), which can hydrogen-bond with solvents and solutes. A comparison of untreated solid supports is shown in Figure IV-2.

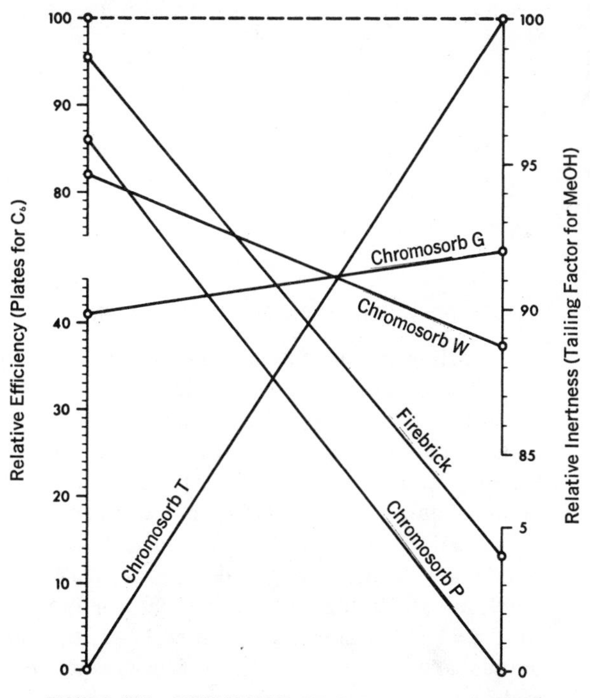

FIGURE IV-2—COMPARISON OF UNTREATED SUPPORTS

Efficiency was obtained from the number of theoretical plates calculated for n-hexane. Efficiency of the Chromosorb T column was the lowest. The number of plates for T was subtracted from each value, making T have the value of 0%. Firebrick with fines removed was the most efficient, and it was assigned a value of 100%. All of the other supports were ranked in % of the difference between T and Firebrick (no fines). Thus in decreasing order of efficiency, we have Firebrick, P, W, G and T.

The *inertness* was determined by measuring the tailing factor for methanol as shown in Figure IV-3.

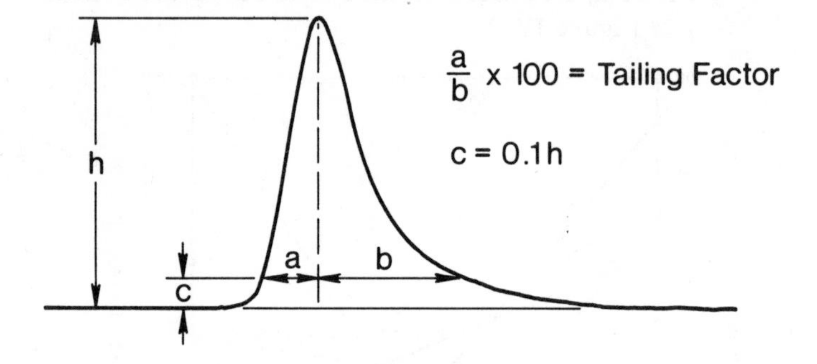

FIGURE IV-3—CALCULATION OF TAILING FACTOR

This tailing factor compares the base of second half of the chromatographic peak to the first half. Base width is measured at 10% of peak height. Two measurements were made for each support--efficiency and inertness (ability to analyze polar compounds). Ten foot columns were prepared of 15% squalane on each of the supports tested. The most inert in decreasing order are T, G, W, firebrick and P. Values of 100% and 0% were assigned to T and P as shown in Figure IV-2; the other values are relative to the difference between T and P.

2. Support surface effects

It is recognized that the diatomaceous supports are not inert and that they differ significantly in adsorption and catalytic activity. Surface interaction can be recognized by tailing and distorted peaks. Catalytic effects can give rise to unexpected peaks.

The hydroxyl and oxide groups on the surface of the support can be masked with a small amount of polar liquid phase. However, the amount masked would vary with the liquid loading and this is difficult to reproduce. A more general procedure is to react the hydroxyls chemically with a reagent. The hydrogen of reactive hydroxyls may be replaced with sily groups from DMCS (dimethyl dichlorosilane) or HMDS (hexamethyldisilizane) as shown in Figures IV-4 and IV-5 respectively.

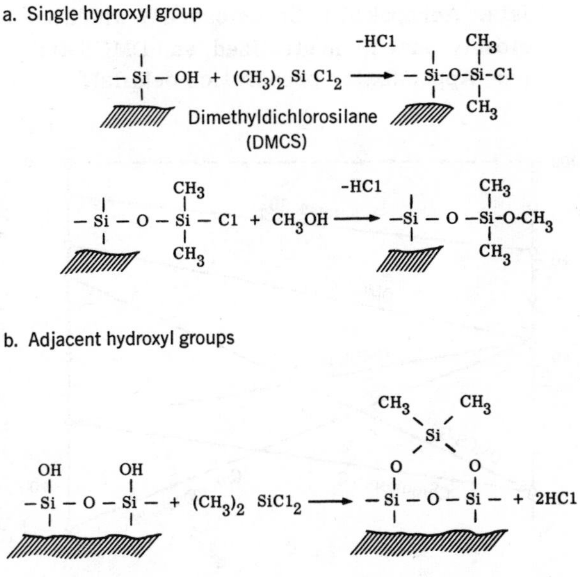

a. Single hydroxyl group

$$-Si-OH + (CH_3)_2SiCl_2 \xrightarrow{-HCl} -Si-O-Si(CH_3)_2-Cl$$

Dimethyldichlorosilane (DMCS)

$$-Si-O-Si(CH_3)_2-Cl + CH_3OH \xrightarrow{-HCl} -Si-O-Si(CH_3)_2-O-CH_3$$

b. Adjacent hydroxyl groups

$$(OH)-Si-O-Si-(OH) + (CH_3)_2SiCl_2 \longrightarrow -Si(O)-O-Si(O)- \text{ [bridged by } Si(CH_3)_2] + 2HCl$$

FIGURE IV-4—SUPPORT REACTIONS WITH DMCS

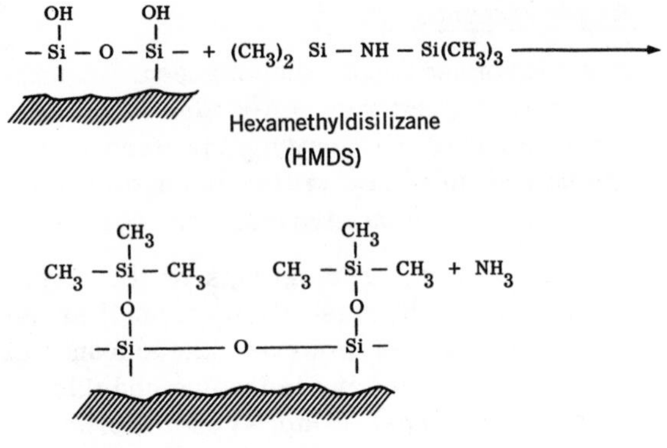

$$-\underset{|}{\overset{\overset{\text{OH}}{|}}{\text{Si}}}-\text{O}-\underset{|}{\overset{\overset{\text{OH}}{|}}{\text{Si}}}- \; + \; (\text{CH}_3)_2\ \text{Si}-\text{NH}-\text{Si}(\text{CH}_3)_3 \longrightarrow$$

Hexamethyldisilizane
(HMDS)

$$\begin{matrix} \text{CH}_3 & & \text{CH}_3 & \\ | & & | & \\ \text{CH}_3-\text{Si}-\text{CH}_3 & & \text{CH}_3-\text{Si}-\text{CH}_3 & + \ \text{NH}_3 \\ | & & | & \\ \text{O} & & \text{O} & \\ | & & | & \\ -\text{Si} & \text{———— O ————} & \text{Si}- & \\ | & & | & \end{matrix}$$

FIGURE IV-5—SUPPORT REACTION WITH HMDS

The effect of these treatments are shown in Figure VI-6. They show Aeropak 30 comes closest to being an ideal support for nonpolar and polar materials. Aeropak 30 is Chromosorb W which has been closely sized, neutralized, and DMCS-treated.[9] (see Appendix for preparation details).

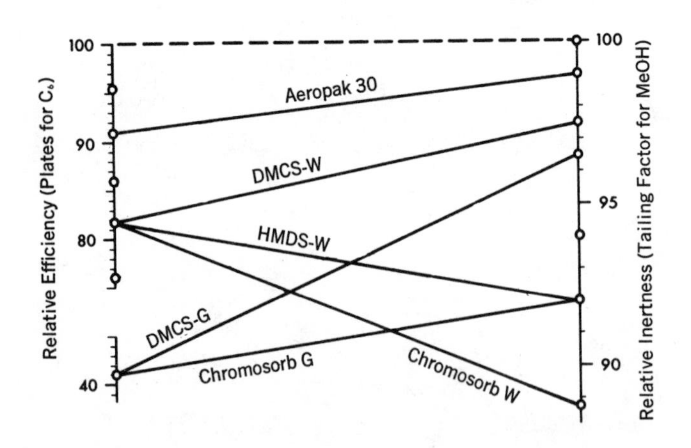

FIGURE IV-6—COMPARISON OF TREATED SUPPORTS

2. Particle size

Referring to the Van Deemter equation (Chapter III) an obvious way to reduce the A term is to reduce dp, the particle diameter. However, as particle diameter is reduced, the pressure drop through the column is increased eventually limiting operation. Column efficiency also improves with the use of narrow mesh ranges. The more uniform the packing, the smaller the "A" term. For 1/8" diameter columns, 100/120 or 80/100 mesh support is preferred; for 1/4" 40/60 on 60/80 mesh.

Supports are usually sized by screening through standard ASTM screens. Mesh numbers refer to the number of openings per linear inch. Particles that will pass through 60 mesh, but not 80 mesh are referred to as 60/80 mesh.

D. LIQUID PHASE PERCENTAGE

The amount of liquid phase used should be enough to coat the particles with a thin uniform layer. Too much liquid phase collects in pools between particles and efficiency of the column decreases. Efficiency decreases drastically for liquid loadings exceeding 30% on diatomaceous earth supports. Several years ago, loadings of 15-30% were common on diatomaceous supports. Now the trend is to lightly-loaded columns (2-10%) and more rapid analyses. On Teflon supports, the maximum load is about 10 wt.%. Glass beads have such a small surface area that liquid loadings should be kept below 0.25 wt. %. The retention time is proportional to the grams of liquid phase present, so low liquid loadings mean fast analyses.

The change of retention time with % liquid loading, is illustrated in Figure IV-7.

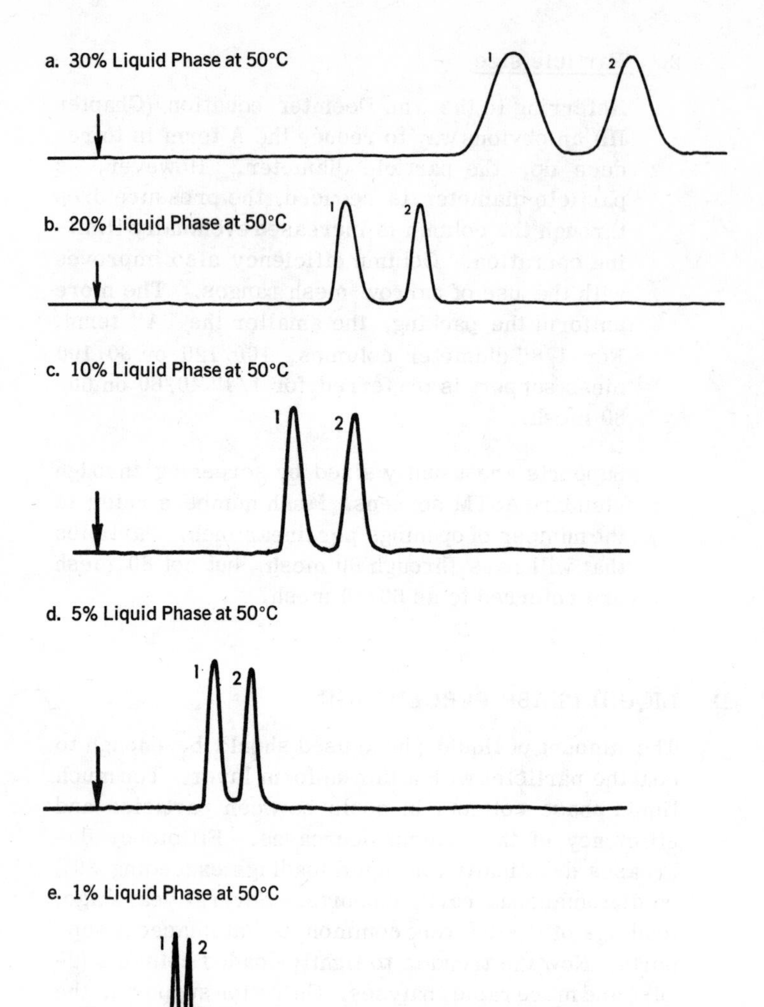

FIGURE IV-7—RETENTION TIME VS. % LIQUID PHASE

Too low a liquid load can leave adsorbing sites on the support exposed. This could cause irreversible adsorption or decomposition of the sample. To avoid this very inert supports such as Aeropak 30 or Gas Chrom P should be used with low liquid loadings. Both Teflon and glass beads are inert, but their efficiency is too low for steroid, pesticide or natural product analyses.

The volatility of the sample should also be considered when choosing the amount of liquid phase. Compounds of low volatility are best run on low-loaded columns --3% or less for materials such as steroids. Very volatile materials, such as light hydrocarbons, require high liquid loads-- 20% to 30%, since their solubilities in the liquid phases are low. The more liquid phase, the longer time which is spent in the liquid, the better the partitioning.

E. COLUMN TEMPERATURE

The partition coefficient is very temperature dependent. In most cases, an increase of 30°C will halve the partition coefficient and thus double the rate of component migration. The most immediate effect of temperature increase is a decrease in analysis time as shown in Figure IV-8 on the following page.

Generally resolution can be improved by lowering the temperature. In the usual case, temperature is chosen by compromise; not so high as to impair resolution, and not so low as to cause long retention times. The temperature is set about equal to the average boiling point of the sample. However, high liquid loadings will require the use of high temperatures, or a highly retentive liquid will require an offsetting temperature increase.

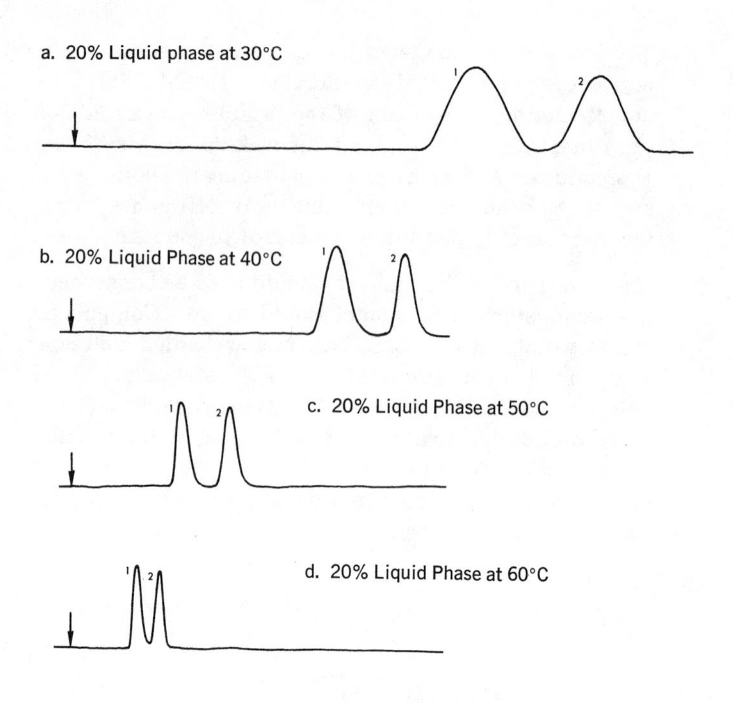

FIGURE IV-8—RETENTION TIME VS. TEMPERATURE

In selecting column temperature, one must also keep in mind the maximum and minimum operating temperatures of the liquid phase being used. Some liquids decompose and cannot be used above a certain temperature; others merely vaporize changing the % liquid phase slowly. A list of maximum recommended temperature of the liquid phases is given in the Appendix, These temperatures should not be exceeded, as this may ruin the column. Some liquid phases do have minimum operating temperatures - Versamid 900 - 150°C, Carbowax 20M and FFAP - 60°C. Operating temperature should be above the melting point of the liquid phase in order for it to function effectively. The Appendix contains instructions for preparing columns.

BIBLIOGRAPHY

1. *Bednas, M.E., and Russell, D.S., Can. J. Chem. 36, 1272 (1958).*
2. *Blake, C.A., Anal. Chem. 35, 1759 (1963).*
3. *Blandenet, G., Robin, J., J. Gas Chromatog. 2, 225 (1964).*
4. *Dewar, M.J.S., and Schroeder, J.P., J. Am. Chem. Soc. 86, 5235 (1964).*
5. *Dewar, M.J.S., and Schroeder, J. P., J. Org. Chem. 30, 3485 (1965).*
6. *Ewell, R.N., Harrison. J.M., and Berg, L., Ind. Eng. Chem. 36, 871 (1944)*
7. *Ferrin. C.R., Chase, J.O., and Hurn, R.W., Gas Chromatography, N. Brenner, et al., eds, Academic Press, New York, 1962, p. 423.*
8. *Gil-Av, E., Herling, J., and Shabtai, J., J. Chromatog. 1, 509 (1958).*
9. *Horning, E.C., Moscatelli, E.A., Sweeley, C.C., Chem. & Ind. (London) 1959, 751.*
10. *Mortimer, J.V., and Gent, P.L., Nature 197, 789 (1963).*
11. *Spencer, S.F., Anal Chem. 35, 592 (1963).*
12. *Van de Craats, F., Anal. Chim. Acta 14, 136 (1956).*
13. *White, D., Nature 179, 1075 (1957).*

V. DETECTORS

A. INTRODUCTION

The chromatographic detector is a device which indicates and measures the amount of seperated components in the carrier gas.

Detectors may be classified as "integrating" or "differentiating". An integrating detector gives a response proportional to the total mass of component in the eluted zone. When pure carrier gas passes through the detector the strip chart shows a straight line. As a component zone passes through, the recorder pen moves across the chart by a distance proportional to the total mass of the component in the zone. When another component zone is eluted the pen moves further across the chart. The chromatogram produced by an integrating detector consists of a series of steps, in which the distance between consecutive level portions of the curve is proportional to the total mass of the component corresponding to that step. The titrating burette is an example of an integrating detector.

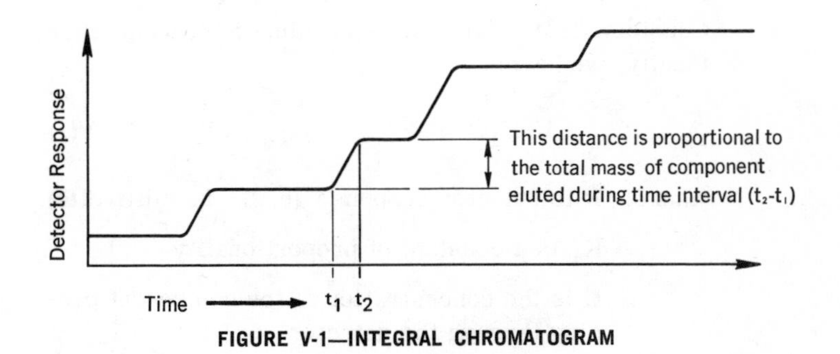

FIGURE V-1—INTEGRAL CHROMATOGRAM

A differentiating detector gives a response proportional to the concentration or mass flow rate of the eluted component. The most familiar example of a detector responding to concentration is the thermal conductivity detector (katharometer). The flame ionization detector is an example of a detector responding to mass flow rate. The chromatogram produced by a differentiating detector consists of a series of peaks, each of which corresponds to a different component. The <u>area</u> under each peak is proportional to the total mass of that component. Differentiating detectors are more commonly used because of their convenience and accuracy.

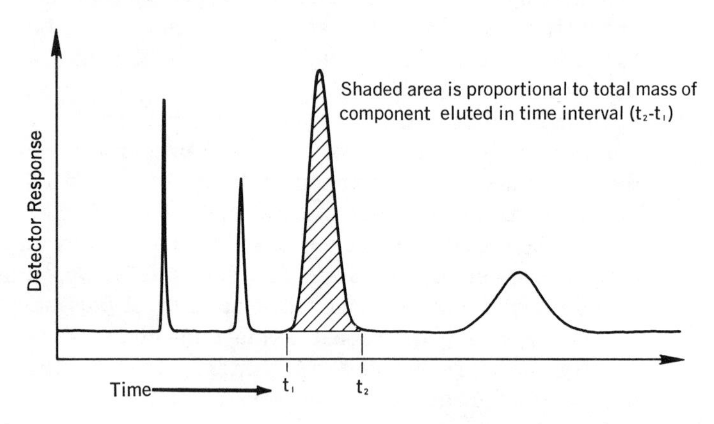

FIGURE V-2—DIFFERENTIAL CHROMATOGRAM

Consider first a detector responding to concentration, ideally, we have:

$$R = K_1 C \tag{1}$$

where: R is detector response (e.g., in millivolts)

K_1 is a constant of proportionality

C is the concentration of the component passing through the detector

If the detector response R is plotted against time t, we get a curve shown in Figure V-3. If the shaded area is A:

$$A = \int_{t_1}^{t_2} R\,dt \tag{2}$$

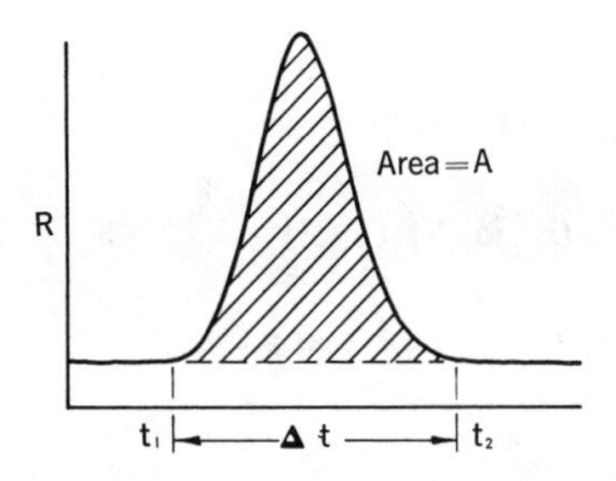

FIGURE V-3—DETECTOR RESPONSE VS. TIME

Substituting the value for R given in equation (1):

$$A = \int_{t_1}^{t_2} K_1 C\,dt = K_1 \int_{t_1}^{t_2} C\,dt \tag{3}$$

Now consider a component zone in the form of a "plug" in which the component concentration is constant (Figure V-4) and equal to M/V. M is the total mass of component in the plug and V is the plug volume. If C is constant:

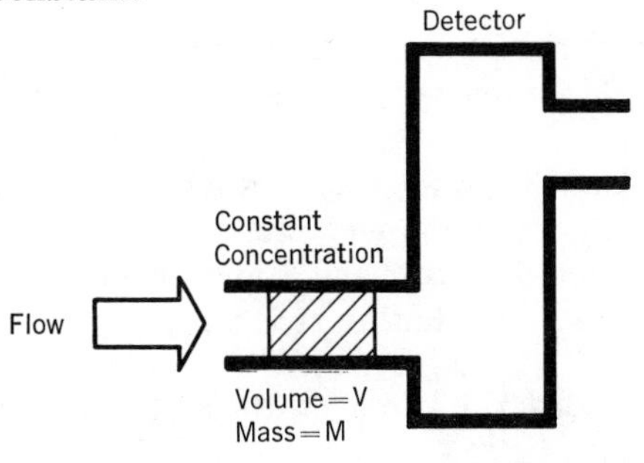

FIGURE V-4—"PLUG" FLOW

$$A = K_1C \int_{t_1}^{t_2} dt = K_1C(t_2 - t_1) = K_1C\Delta t \quad (4)$$

But $C = M/V$, therefore:

$$A = (K_1M/V)\,\Delta t \quad (5)$$

However, $V = F \cdot t$, where F is carrier gas flow rate. Thus, $\Delta t = V/F$, and:

$$A = (K_1M/V)\,(V/F) = K_1M/F \quad (6)$$

Thus we obtain the desired result:

$$A = K_1M/F \quad (7)$$

Equation (7) shows that peak area is directly proportional to the total mass of component. It is this relation that enables the chromatographer to calculate weight percent compositions from area ratios on the chromatogram. One must know the proper correction factor, F, (see Chapter VII). It is important to note that equation (7) also shows that, for a detector <u>responding</u> to concentration, the peak area is inversely proportional to carrier gas flow rate. Thus, for accurate quantitative analysis with thermal conductivity detectors, the flow rate must be kept constant.

For detectors responding to mass flow rate (dm/dt), such as the flame ionization detector, we have:

$R = K_2(dm/dt)$, where K_2 is a new constant of (8) proportionality, m is the instantaneous mass of component within the detector, and R and t have their previous meanings. By similar reasoning:

$$A = \int_{t_1}^{t_2} R\,dt = \int_{t_1}^{t_2} K_2\,(dm/dt)\,dt = K_2 \int_{t_1}^{t_2} (dm/dt)\,dt \quad (9)$$

Cancelling dt within the integral sign, and integrating, we obtain:

$$A = K_2M \qquad (10)$$

Equation (10) shows that for a detector responding to mass flow rate, the peak area is proportional to the total mass of the eluted component. However, unlike the concentration detector, the peak area for a mass flow detector is independent of carrier gas flow rate. Constant flow rate is not as critical for an FID as for a TC detector.

B. DETECTOR CHARACTERISTICS

Since chromatographic detectors differ greatly in the principal on which they operate, it is difficult to compare them. Certain characteristics, however, are indicative of the usefulness of the detector:

1. Sensitivity
2. Noise level
3. Linear range
4. Response

In addition, there are secondary characteristics which should be considered. If possible, the detector should be simple, inexpensive, rugged and insensitive to changes in flow rate and temperature. A universal detector should respond to all types of components. There is also a need for specific detectors such as the electron capture or phosphorus detectors which respond selectively to only certain classes of compounds.

1. Sensitivity

For detectors responding to concentration, sensitivity is the detector response R, usually millivolts, per unit concentration of the component. Dimbat, Porter, and Stross [1] have defined a unit of sensitivity equivalent to an output of 1 millivolt per milligram of component per cubic centimeter of carrier gas. The sensitivity of the detector can then be expressed:

$$S = mv/(mg/cm^3) = (mv \cdot cm^3)/mg$$

The sensitivities of thermal conductivity cells are about 1000 to 10,000 $(mv \cdot cm^3)/mg$. In addition to specifying an appropriate unit for detector sensitivity, Dimbat, Porter, and Stross have also derived an expression for determining sensitivity in terms of easily measured parameters:

$$S = (A \bullet c_1 \bullet c_2 \bullet c_3)/w \qquad (11)$$

where: S = detector sensitivity in $(mv \cdot cm^3)/mg$
A = peak area in cm^2
c_1 = recorder sensitivity in mv/cm of chart
c_2 = reciprocal of chart speed in min/cm
c_3 = flow rate of carrier gas in ml/min
w = weight of component in mg.

For detectors responding to mass flow rate, the Dimbat sensitivity unit must be modified to:

$$S' = mv/(mg/sec) \quad (mv \cdot sec)/mg$$

Similarly, the sensitivity of a mass flow detector is given by:

$$S' = (A \bullet c_1 \bullet c_2)/w \qquad (12)$$

where: S' = detector sensitivity in $(mv \cdot sec)/mg$
c_1 = recorder sensitivity in mv/cm of chart
c_2 = reciprocal of chart speed in sec/cm

A = peak area in cm^2
w = weight of component in mg

Note that S' is independent of flow rate, whereas S is proportional to flow rate.

2. Noise level

The electrical output of a detector can be increased to almost any desired value by electronic amplification. Thus detector sensitivity can be made as large as desired. However, electrical noise inherent in the detector and electronics is also amplified and a point is reached where the noise level is high enough to hide the detector's response. Therefore, the noise level limits the concentration (or mass flow rate) of components that can be detected.

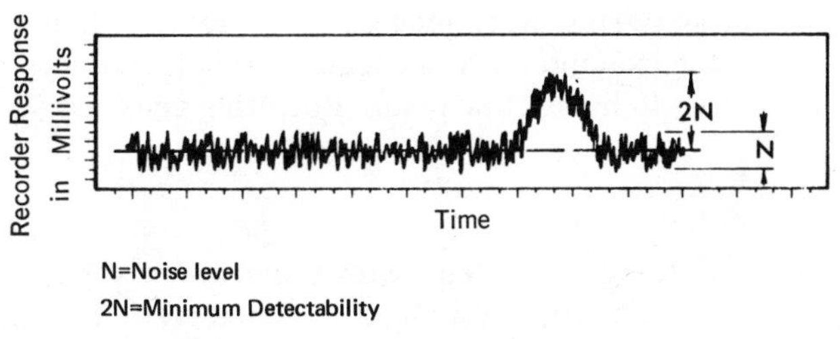

N=Noise level
2N=Minimum Detectability

FIGURE V-5—MINIMUM DETECTABLE QUANTITY

Figure V-5 shows a plot of recorder trace of detector response. The random high frequency spikes represent electrical noise in the amplified signal; the average peak-to-peak value of the noise spikes has been labelled "N". It is evident that a component must give a detector response greater than N in order to be distinguished from background noise. The minimum detectable concentration is that concentration which gives a detector response equal to twice the noise level. For example, if the peak-to-peak noise level is 4 microvolts, then the minimum detectable concentration is that which gives a detector response of 8 microvolts. Many chromatographers use the minimum detectable concentration as a measure of detector sensitivity rather than expressing sensitivity in terms of $(mv \bullet cm^3)/mg$. This usage is practical since it is the detector noise level which ultimately determines the lower limit of detectability. A detector might be extremely sensitive, but if it were also extremely noisy, it would not be useful sensitivity.

3. Linear range

Accurate quantitative analysis depends upon a linear relation between concentration and detector response. The more closely a linear relation is obeyed the more accurate the analysis. Consider, for example, a flame ionization detector responding to mass flow rate. Recalling equation (8):

$$R = K_2 \, (dm/dt) \tag{8}$$

If R were plotted against dm/dt we would get a straight line with slope K_2. However, the ranges of R and dm/dt are usually so wide in practice that it is more convenient to plot equation (8) on a log-log scale. Taking logarithms of both sides of equation (8), we obtain:

$$\log R = \log K_2 + \log \, (dm/dt) \tag{13}$$

Equation (13) is of the form $y = a + bx$, where $b = 1$. Thus, a plot of equation (8) on a log-log scale should give a straight line having a slope of 1. The response index or "linearity" of a detector may be defined as the slope of the detector response curve plotted on log-log scale. Thus, a perfectly linear detector would have a slope of 1.00. In practice flame ionization detectors have linearities in the range 0.95 to 0.99.

The "linear range" of a detector may be defined as the ratio of the largest to the smallest concentration within which the detector is linear. The term "linear dynamic range" (LDR) is sometimes used synonymously for "linear range".

To illustrate the terms "linearity" and "linear range", Figure V-6 shows a log-log plot of actual data obtained with a flame ionization detector. Both detector response and mass flow rate are in arbitrary units.

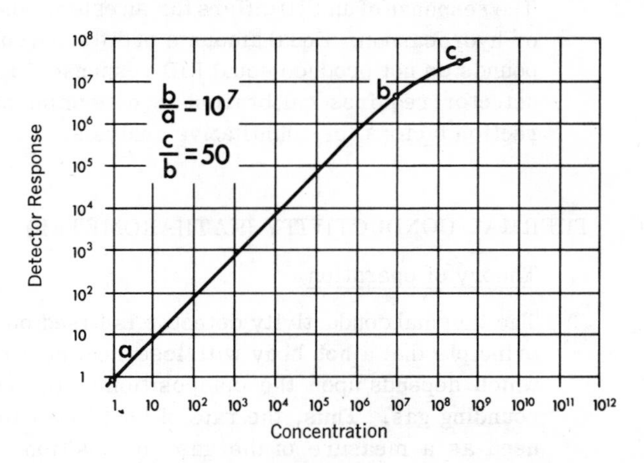

FIGURE V-6—LINEARITY OF A F.I.D.

The measured slope between points a and b is 0.97; hence, by definition, the detector linearity is 0.97. Point b represents the highest mass flow rate for which the linearity (slope) was between 0.95 and 1.00 (5% was arbitrarily selected as the tolerance on linearity). Beyond point b linearity falls below 0.95. The linear range is, by definition, b/a, or 10^7. In practical terms, one could use the same calibration curve for component concentration changes of ten million fold.

4. Response

The response of a detector depends upon the operating principle. A TC detector responds to the change in thermal conductivity between the sample and carrier gas. Thermal conductivity is proportional to molecular weight for many compounds. This means the detector response changes for compounds of different molecular weights and correction factors (see Chapter VII) must be used.

The response of an FID differs for an ester, ether, or hydrocarbon. Equal amounts of different compounds do not produce equal FID response. Each detector requires calibration to determine correction factors for quantitative analysis.

C. THERMAL CONDUCTIVITY (KATHAROMETER)

1. Theory of operation

The thermal conductivity detector is based on the principle that a hot body will lose heat at a rate which depends upon the composition of the surrounding gas. Thus, the rate of heat loss can be used as a measure of the gas composition. An early apparatus for determining the purity of gas streams was patented in 1915 by Shakespear and called a "katharometer" (from the Greek word "katharos", meaning pure). The TC detector was introduced into gas chromatography by Claesson[10] in 1946 and has remained a major detector ever since.

Figure V-7 shows a typical TC cell consisting of a spiral metal filament supported inside a cavity within a metal block.

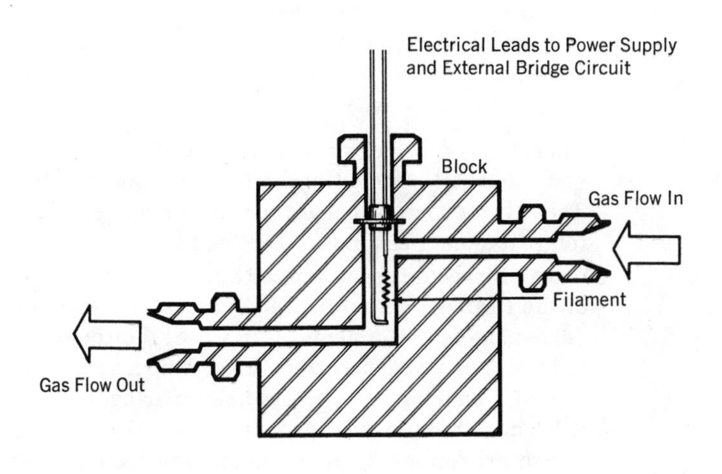

FIGURE V-7—SCHEMATIC OF T.C. CELL

The heated filament can lose heat by the following processes:

1. Thermal conduction to the gas stream
2. Convection (free and forced)
3. Radiation
4. Conduction through the metal contacts

Heat conduction through the metal filament contacts is negligible because of the small contact area. Heat loss by radiation is proportional to $(T_f^4 - T_b^4)$, where T_f is the absolute temperature of the filament and T_b is the absolute temperature of the block. Calculations made for a typical filament temperature of 400°C, block temperature of 300°C, filament diameter of 0.001 cm, and filament total length (uncoiled) of 10 cm show that the radiation heat loss is about 10^{-6} calories per second, which is negligible. Free convection is also negligible because of the small internal diameter of the filament cavity.

Therefore, the major heat loss processes are gaseous thermal conduction and forced convection. These two process account for 75% or more of the total filament heat loss. The heat loss by forced convection could be minimized by proper geometry of the filament within the block cavity (so-called "diffusion-fed" filaments). However, diffusion-fed cells have undesirably long response time. Use of a carrier gas such as helium or hydrogen, will cause heat loss by gaseous thermal conduction to predominate. It is assumed in the following discussion that thermal conduction by the carrier gas is the only mode of heat transfer.

Heat is transferred by conduction when gas molecules strike the heated filament and rebound with increased kinetic energy. The greater the number of molecular collisions with the filament per unit time, the greater the rate of heat loss. Differences in thermal conductivity of gases are based on the mobility or speed at which the gas molecules can diffuse to and from the hot filament. The speed of molecules is a function of molecular weight, and as a result, the smaller the molecule the higher its speed and the higher its thermal conductivity. Thus, hydrogen and helium which are the smallest molecules have the highest thermal conductivity. Table V-1 gives the thermal conductivities (in c.g.s. units and at 0°C) of several common gases and shows how thermal conductivity decreases with increasing molecular weight.

TABLE V-1—THERMAL CONDUCTIVITY

	λ x 10^5	Molecular weight
Hydrogen	41.6	2
Helium	34.8	4
Methane	7.2	16
Nitrogen	5.8	28
Pentane	3.1	72
Hexane	3.0	86

2. TC sensing elements

A TC Cell consists of a spiral filament wire, supported inside a cavity within a metal block. The filament is made of a material whose electrical resistance varies greatly with temperature, i.e., it has a high temperature coefficient of resistance. A constant current is passed through the filament causing its temperature to rise. In a typical TC cell, with helium carrier gas and a filament current of 175 milliamperes, the filament may reach a temperature of 100°C above the block temperature. The filament temperature is determined by the equilibrium between the electrical power input (I^2R) and the thermal power loss due to heat conduction by the surrounding gas. With pure carrier gas flowing, the heat loss is constant and thus the filament temperature is also constant.

If the gas composition changes, e.g., when a sample peak emerges, the filament temperature changes, causing a corresponding change in electrical resistance. It is this resistance change which is measured by a Wheatstone bridge circuit.

Filament metals are chosen on the basis of high temperature coefficient of resistance, and resistance to chemical corrosion. Common filament metals are platinum, tungsten, and tungsten alloys. Table V-2 shows the characteristics of several commonly used filaments.

TABLE V-2—FILAMENT CHARACTERISTICS

Gow-Mac Filament	Ohms (cold)	Relative Response	Maximum Recommended Current N_2	He
W(Tungsten)	20	1.0	175 ma	350 ma
WX(Tungsten + 3% Rhenium)	31	2.0	150	300
W-2 (Tungsten: Two spiral)	40	2.2	125	300

Some TC cells use thermistors rather than metal filaments. Thermistors are sintered mixtures of manganese, cobalt, and nickel oxides, plus trace elements to give the desired electrical properties. Thermistors have a very high negative coefficient of resistance as contrasted to metals. The thermistor, in the form of a small bead, is mounted on platinum wire and coated with glass to make it inert. Thermistors are very sensitive but have a limited temperature range and poor stability. Their sensitivity decreases with increasing temperature, and they are usually employed at room temperature. For more information on thermistors, see Dal Nogare[17]

3. Electrical circuit

The change in filament resistance must be measired and converted to an output signal. Figure V-8 shows a simple Wheatstone bridge circuit. When all four filaments S_1, S_2, R_1, and R_2 are at the same temperature, and hence have the same resistance, the bridge is balanced and there is zero output. However, if the resistance of filaments S_1 and S_2 changes due to a change in gas composition, an unbalance of the bridge occurs and an output signal is generated. Most detector blocks contain a pair of matched filaments (S_1 and S_2) in the sample flow channel, and a similar pair of matched filaments (R_1 and R_2) in the reference flow channel. This arrangement provides twice the output signal of a two filament bridges, as well as stabilizing the bridge against ambient temperature fluctuations. The reference and sample flow channels are drilled into a single metal block having a high heat capacity. This provides temperature stabilization.

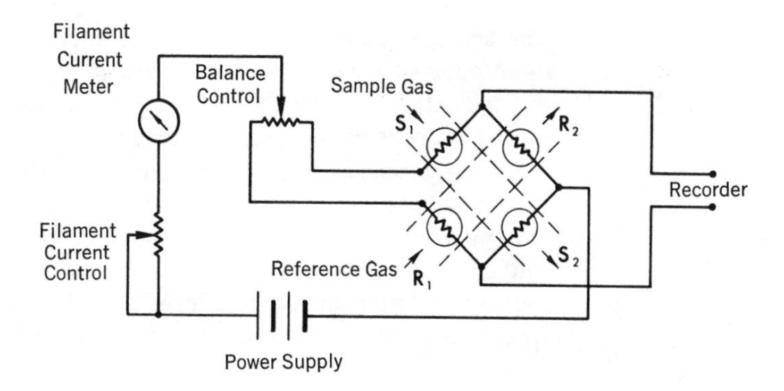

FIGURE V-8—T.C. WHEATSTONE BRIDGE CIRCUIT

4. Factors affecting sensitivity

Equation (17) gives the sensitivity of a TC cell in terms of the cell parameters:

$$(17)\quad S = K \bullet I^2 \bullet R \frac{(\lambda c - \lambda s)}{\lambda c} (T_f - T_b)$$

where:
- S = sensitivity
- K = cell constant dependent on geometry
- I = filament current
- R = filament resistance
- λc = thermal conductivity of carrier gas
- λs = thermal conductivity of sample gas
- T_f = temperature of filament
- T_b = temperature of detector block

From an examination of equation (17) it becomes apparent how the following factors affect the detector sensitivity:

a. Current

The I^2 term shows that increasing the filament current will increase the output signal significantly. In addition, an increase in filament current corresponds to an increase in filament temperature, and an increase in filament resistance R. The net result is a four- to eight-fold increase in sensitivity for a two-fold increase in filament current. Of course, increasing current too high produces baseline instability and may burn out the filament.

b. Carrier gas

The term $\frac{(\lambda c - \lambda s)}{\lambda_c}$ increases as λ_c gets larger (this can easily be shown by inserting a few numbers for λ_c while keeping λ_s constant). Therefore, choose a carrier gas having the highest possible thermal conductivity. Hydrogen or helium provide the highest sensitivity for organic compounds.

c. Temperature

Increasing the filament temperature T_f (by increasing filament current I) will increase detector sensitivity. However, the block temperature should be kept as low as possible, so as to maximize the difference (T_f-T_b). The block temperature should be high enough to avoid condensation of the sample within the detector.

d. Summary

To increase the sensitivity of a TC detector, one should increase the filament current, decrease the block temperature and choose a carrier gas having high thermal conductivity.

5. Operating suggestions for TC detectors

a. Always be sure that carrier gas is flowing through the detector before turning on the filament current. It is very easy to burn out the filaments unless a gas stream is present to dissipate heat.

b. Turn off the filament current before changing columns, putting in new injector septa, or otherwise opening the flow system to the atmosphere. Small volumes of air leaking into the system can oxidize and ruin the heated filaments.

c. Excessive noise, baseline drift, or inability to balance the TC bridge may be caused by corrosion of the filaments. If filaments are badly corroded they must be replaced. However, if baseline starts to drift (after being straight), immediately turn off filament current and check the system for leaks. Sudden baseline drift may be caused by air getting into system through a leak and starting to oxidize the filaments. Air can diffuse into the system even against the positive pressure of carrier gas. If the filament current is turned off immediately, and leak located, the filaments may still be in usable condition.

d. Excessive noise and baseline drift may also be caused by high boiling components which have condensed on the filaments. Cool the detector block to room temperature, disconnect the column, and place an injector septum and nut on the detector inlet. Inject sufficient solvent (benzene or xylene) to fill the flow channels, let stand overnight. Hot xylene is a good solvent for silicone polymers. Clean and dry thoroughly before using.

e. Samples such as HCL, chlorine, fluorine, alkyl halides, organofluorides, and other reactive compounds will readily damage standard TC filaments. If working with such compounds, use a nickel detector block equipped with Teflon-coated filaments. There is a two to three fold loss in sensitivity.

f. TC detectors are flow sensitive. The carrier flow rate should be held constant by means of a two-stage regulator. Temperature programming with a TC detector requires a differential flow controller because of the expansion of the carrier gas with increasing temperature. High tank pressure must be used to insure baseline stability while programming.

THERMAL CONDUCTIVITY DETECTOR		
Min. Det. Quant.	-	2 - 5 ug (100 ppm in 25 ul liquid or 100 ppm in 5 ml gas.)
Response	-	All components except carrier gas
Linearity	-	10,000
Stability	-	Good
Carrier Gas	-	Helium, hydrogen, nitrogen
Temperature Limit	-	450°C
Summary - Nondestructive, stable, moderate sensitivity, inexpensive, simple to operate. Requires good temperature and flow control.		

D. IONIZATION DETECTORS

1. General theory of operation

Ionization detectors operate on the principle that the electrical conductivity of a gas is directly proportional to the concentration of charged particles within the gas. Figure V-9 shows a generalized ionization detector in which the ionizing source is unspecified.

Effluent gas from the column flows through the electrode gap past an ionizing source which ionizes some of the molecules in the gas stream. The presence of charged particles (positive ions, negative ions, electrons) within the electrode gap causes a current, I, to flow across the gap and through a measuring resistor R_2. The resulting voltage drop E_o across R_2 is amplified by an electrometer and fed into a recorder.

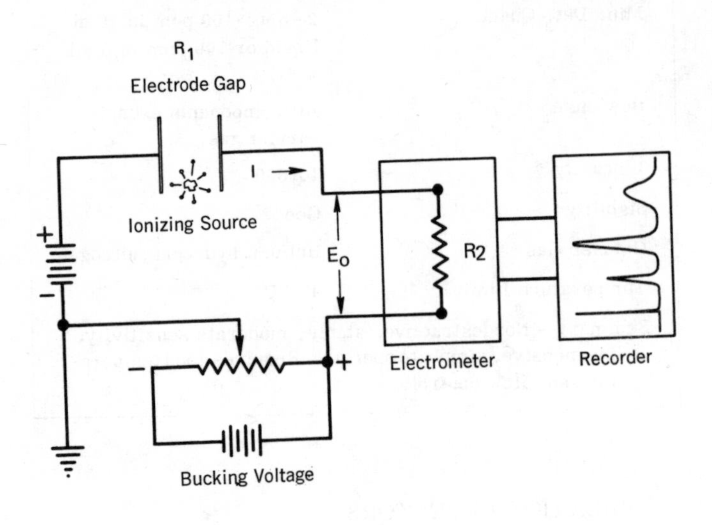

FIGURE V-9—SCHEMATIC IONIZATION DETECTOR CIRCUIT

It is helpful to think of the electrode gap as a variable resistor R_1 whose resistance value is determined by the number of charged particles within the gap. With pure carrier gas flowing a constant concentration of charged particles will be present in the gap, thus causing a constant current, I, to flow. This constant current is called

the "background current". It is desirable to minimize the background current so small changes in current can be more easily measured (it is easier to measure a fixed change in a small quantity than the same change in a larger quantity). The background current is reduced to zero by opposing it with a "bucking voltage". Thus, under "no signal" conditions (pure carrier gas flow), no current flows, and the recorder traces a straight baseline. When a sample component passes through the electrode gap, molecules of the component are ionized. This increases the number of charged particles and decreases the value of R_1.
This decrease permits current to flow which produces a signal which is registered as a peak on the recorder.

2. Flame ionization detector

a. Operating principle

Figure V-10 shows a typical FID in which the effluent gas from the column is mixed with hydrogen and burned in air or oxygen. Ions and electrons formed in the flame enter the electrode gap, decrease the gap resistance, thus permitting a current to flow in the external circuit. When the FID was first introduced in 1958, it was assumed that thermal ionization was the mechanism operating. Recent evidence indicates that thermal ionization may play only a minor role in the overall ionization mechanism. Sternberg et al [(2)] give a comprehensive discussion of flame ionization theories.

b. Detector response

The FID responds to virtually all compounds with the exception of those listed in Table V-3; note particularly the lack of response to air,

H_2O and CS_2. The lack of response to air and water makes the FID especially suitable for the analysis of air pollutants or aqueous samples such as alcoholic beverages, biological materials, etc. Similarly, the absence of a "solvent peak" makes carbon disulfide a convenient solvent for use with the FID.

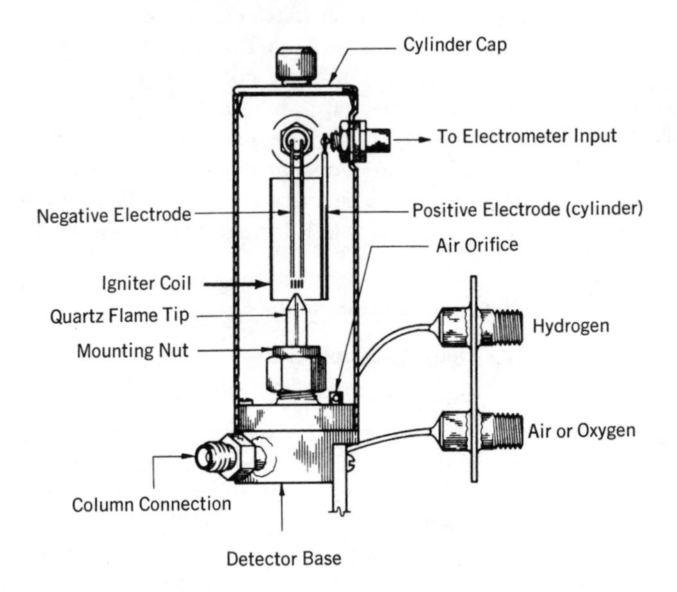

FIGURE V-10—FLAME IONIZATION DETECTOR

Accurate quantitative analysis with the FID can only be done by using individual response factors for each component of interest. The calculation of these response factors and specific values for the FID are given in Chapter VII on Quantitative Analysis.

TABLE V-3—COMPOUNDS GIVING LITTLE OR NO RESPONSE IN THE FLAME IONIZATION DETECTOR

He	CS_2	NH_3
Ar	COS	CO
Kr	H_2S	CO_2
Ne	SO_2	H_2O
Xe	NO	$SiCl_4$
O_2	N_2O	$SiHCl_3$
N_2	NO_2	SiF_4

TABLE V-4—QUANTITATIVE ANALYSIS WITH FLAME IONIZATION

c. Flow rate vs detector response

The FID performance depends on the proper choice of gas flow rates. Use a 1:1:10 relative flow ratio of hydrogen/carrier/air. Thus a carrier flow of 30 ml/min requires 30 ml/min of hydrogen and 300 ml/min of air for optimum performance of the FID. It is necessary to place an upper limit on the carrier flow rate, otherwise the high linear velocity of the fuel gas (carrier + hydrogen) through the burner orifice may cause the flame to flicker (causing noisy baseline) or actually blow out. Maximum carrier flow for a 1/8 inch O.D. column is about 30 ml/min, and for a 1/4 inch O.D. column about 80 ml/min. It is necessary to use a wider orifice with the higher flow rate in order to reduce the linear gas velocity through the burner.

Figure V-11 shows the relation between FID response (in coulombs per mole) and hydrogen flow rate (ml/min). Note that while the response is different for each compound, the curves have the same shape and all reach a maximum at about 30 ml/min hydrogen flow. The carrier gas flow rate was 30 ml/min for the curves shown. Figure V-11 shows that FID response is a sensitive function of hydrogen flow rate, and that individual detectors should be calibrated to determine the hydrogen flow rate that gives maximum response.

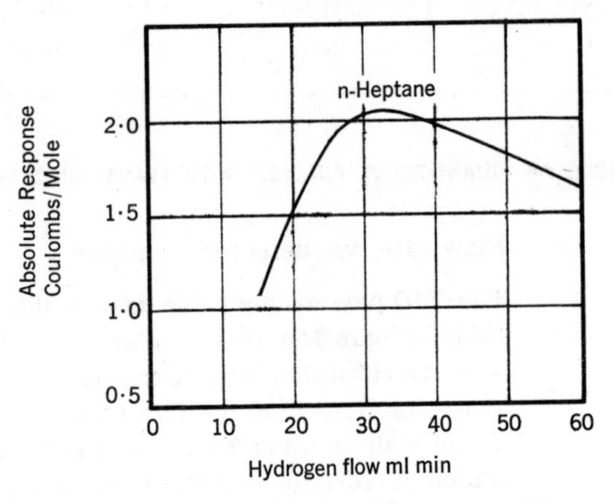

FIGURE V-11—F.I.D. SENSITIVITY VS. H_2 FLOW

Figure V-12 shows the relation between FID response and air flow rate.

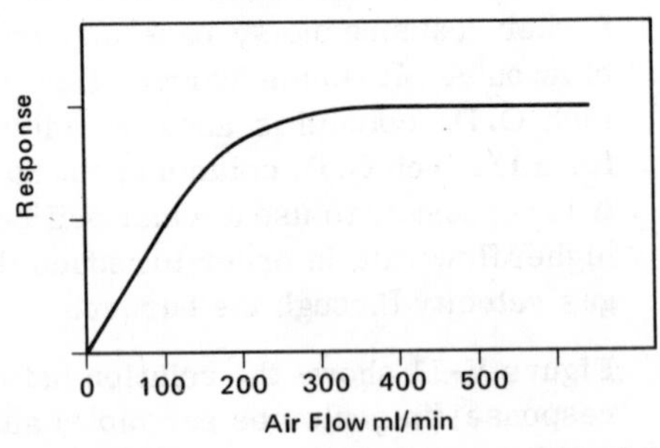

FIGURE V-12–F.I.D. SENSITIVITY vs AIR FLOW

d. Minimum detectable quantity

For a specific set of operating conditions the FID response for n-propane was 30 microamps/(mg/sec). The observed noise level was 5×10^{-8} microamps (μa). Dividing the FID response into twice the noise level, we obtain:

$$2(5 \times 10^{-8})\mu a/30\mu a/mg/sec = 3 \times 10^{-12} gm/sec$$

This is a typical value for FID minimum detectable quantity, the normal range being between 1 and 10×10^{-12} grams/sec. Note that minimum detectable quantity is given in units of mass flow rate rather than mass alone. This is because the FID response is proportional to mass flow rate. Values quoted in terms of mass flow rate can be converted to mass units by multiplying by the peak width at the baseline (in seconds).

FLAME IONIZATION DETECTOR	
Minimum Det. Quant.	- 1 to 10 x 10^{-12} gm/sec.
Response	- Sensitive to organic compounds. not to fixed gases or water
Linearity	- 10^6 to 10^7
Stability	- Excellent (relatively insensitive to temperature and flow changes).
Temperature Limit	- 400°C
Carrier Gas	- Nitrogen or helium
Summary:	Inexpensive, simple, extremely rugged, and can withstand extreme temperature change. Insensitive to fixed gases. Insensitive to water making it especially useful for analyzing dilute aqueous solutions.

e. Linear range

The FID has the widest linear range of any detector in common use. Linear range is between 10^6 and 10^7 (see Figure V-6 for an actual FID linearity plot). The combination of high sensitivity and wide linear range make the FID the choice in trace analysis.

3. Electron capture detector

a. Operating principles

The electron capture detector measures the loss of signal rather than a positively produced electrical current. As the nitrogen carrier gas flows through the detector, a tritium source ionizes the nitrogen molecules and slow electrons are formed. These slow electrons migrate to the anode under the fixed voltage which is termed "cell voltage". Collected, these slow electrons produce a steady current amplified by the electrometer. If a sample containing electron absorbing molecules is then introduced, this current will be reduced. The loss of current is a measure of the amount and electron affinity of the compound.

b. Detector response

The electron capture detector is extremely sensitive to certain molecules, such as alkyl halides, conjugated carbonyls, nitriles, nitrates, and organometals, but is virtually insensitive to hydrocarbons, alcohols, ketones, etc. Selective sensitivity to halides makes this detector especially valuable for the analysis of pesticides. Recent studies at our laboratory and others has shown that certain pesticides can be detected down to the subpicogram quantity (10^{-13} gms).

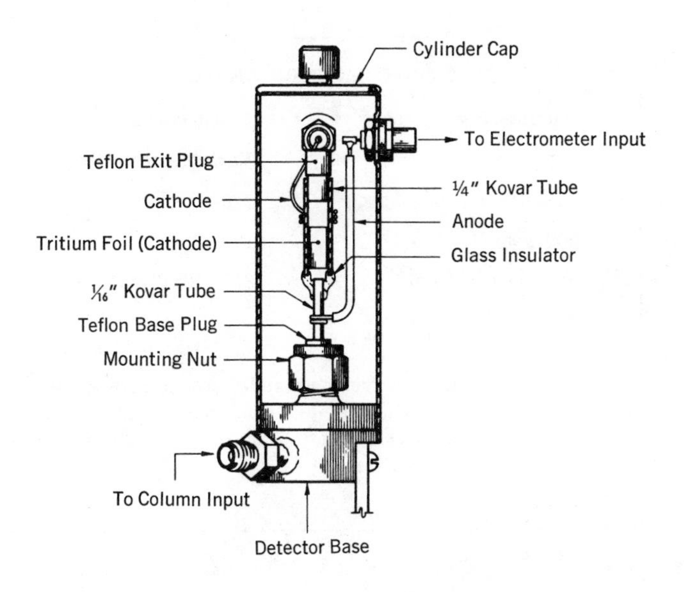

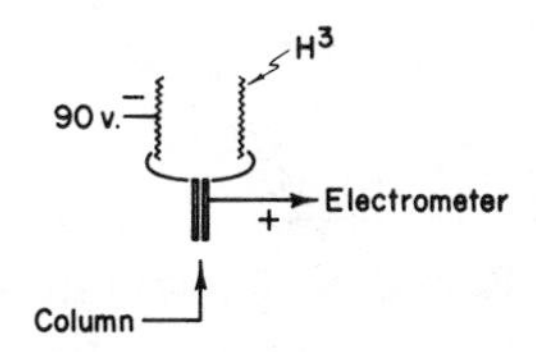

$\beta + N_2 \rightarrow N_2^+ + e^-_{Slow}$

$e^-_{Slow} + X \rightarrow X^-$

Loss of e^- $\rightarrow$ Reduces Current

FIGURE V-12—ELECTRON CAPTURE DETECTOR

ELECTRON CAPTURE DETECTOR		
Minimum Det. Quantity	-	0.0000001 ug
Response	-	Variable sensitivity
Linearity	-	500
Stability	-	Fair
Temperature Limit	-	225^oC (tritium) 350 for Ni^{63}
Carrier Gas	-	N_2 or Argon + 10% CH_4 (with pulsed voltage)

Summary - Inexpensive and simple. Detector is easily contaminated and easy to clean. Sensitive to water; carrier gas must be dry. Can be operated in pulsed or D.C. mode (3).

4. Helium detector (4)

The Helium detector was developed for the trace analysis of permanent gases. It employs the cross-section detector geometry, 400 v electrode potential, and helium as carrier gas.

The combination of the tritium beta radiation and high field gradient (4000 volts per cm) raise the He to a metastable state with an ionization potential of 19.8 eV. All compounds having a lower ionization potential will be ionized giving rise to a positive signal.

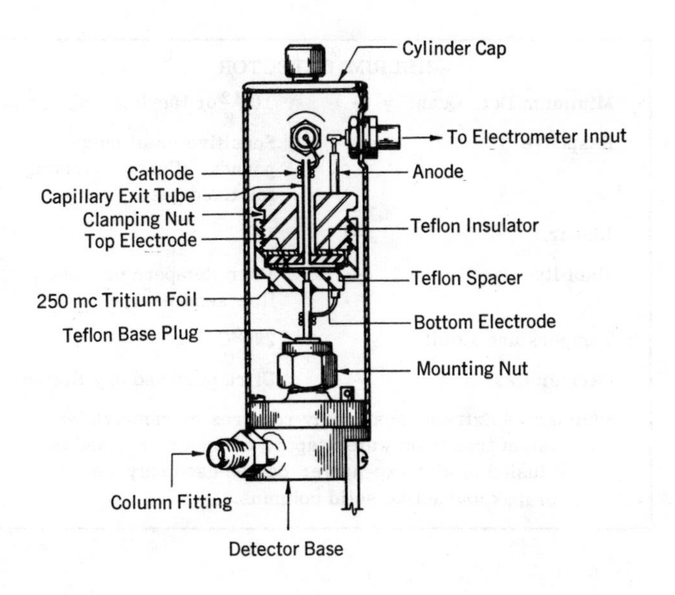
Cylinder Cap
To Electrometer Input
Cathode
Anode
Capillary Exit Tube
Clamping Nut
Teflon Insulator
Top Electrode
Teflon Spacer
250 mc Tritium Foil
Bottom Electrode
Teflon Base Plug
Mounting Nut
Column Fitting
Detector Base

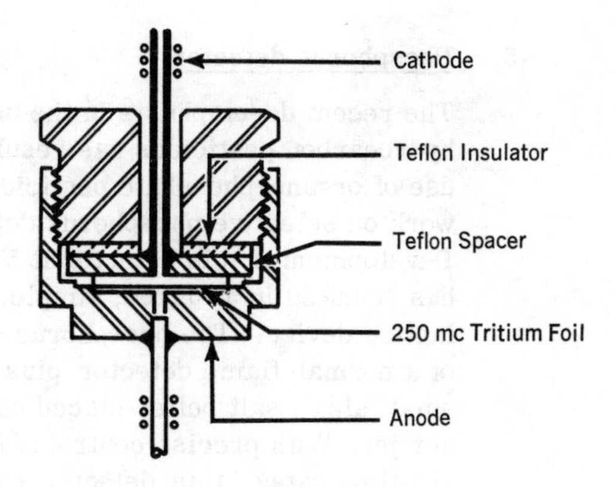
Cathode
Teflon Insulator
Teflon Spacer
250 mc Tritium Foil
Anode

FIGURE V-13—HELIUM DETECTOR

HELIUM DETECTOR		
Minimum Det. Quantity	-	10^{-12} g or 10ppb of 3 mL sample
Response	-	Sensitive to all compounds. Useful primarily for fixed gases.
Linearity	-	10^4
Stability	-	Poor (temperature and flow sensitive)
Temperature Limit	-	225°C
Carrier Gas	-	Ultra pure and dry Helium

Summary - Extreme sensitivity requires extremely clean system free from water vapor, column bleed, and back diffusion of air; expensive; can be used only with Porapak and active solid columns.

5. Phosphorus detector

The recent de-emphasis in the use of chlorinated hydrocarbon pesticides has resulted in increased use of organo-phosphate pesticides. This led to work on selective phosphorus detectors (5, 6). Development of this detector at Varian Aerograph has resulted in a rugged, simple, stable, and long lasting device. The phosphorus detector consists of a normal flame detector plus the addition of a small alkali salt pellet placed on the quartz burner jet. With precise control of the hydrogen and air flow rates, this detector can be made to be very sensitive to phosphorus containing compounds and completely insensitive to other organic materials.

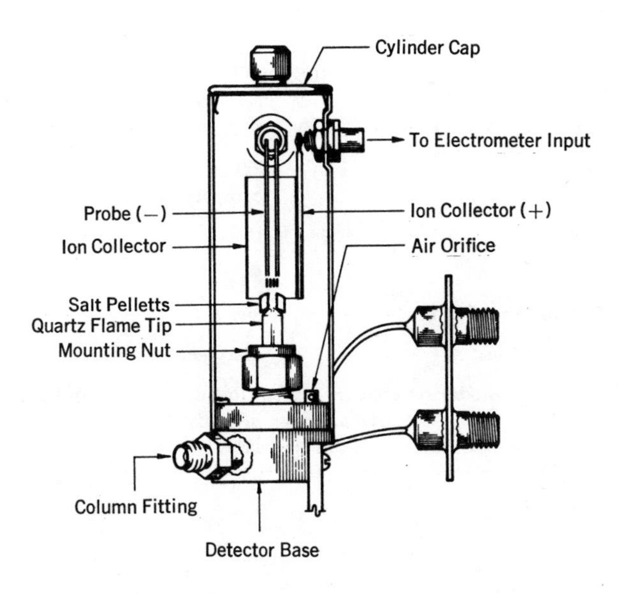

FIGURE V-14—PHOSPHORUS DETECTOR

This modification resulted in a selective detector whose phosphorous response relative to the flame ionization detector is approximately 3000:1.

PHOSPHORUS DETECTOR		
Minimum Det. Quantity	-	0.00001 ug (parathion)
Response	-	Sensitive to phosphorous compounds.
Linearity	-	10^4
Stability	-	Fair
Temperature Limit	-	300°C
Carrier Gas	-	Nitrogen or Helium
Summary:		Moderate cost (requires flow controller for H_2 and air), selective and high sensitivity; destructive, fair stability, good linearity.

6. Cross-section detector

This detector was invented by Shell Development Company about 1957. The design was such that it could be used only with large gas volumes. It was comparatively insensitive but extremely linear with a dynamic range extending from minimum detectable quantities to 90% concentration of the sample.

The response of this detector can be calculated from a chemical basis, i.e., knowing the cross sections of the individual atoms that comprise the sample molecule and the amount injected, the response can be calculated [7,8,9].

The beta particles which emanate from the 250 mc tritium foil have a relatively low probability of collision with the orbital electrons of the carrier gas molecules (H_2 or He), i.e., the cross section for ionization of H_2 and He is very low. The electrons generated by this collision are instantly collected at the anode to produce a small background current. Any other molecule entering the detector will have a higher density of orbital electrons thereby creating a higher probability for beta particle collision. An increase in current will be recorded.

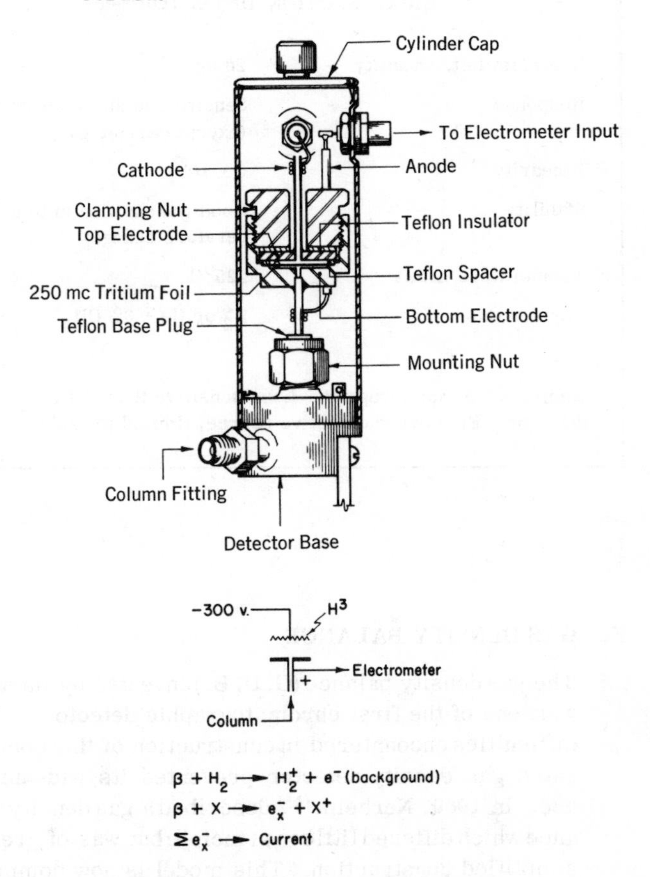

FIGURE V-15—CROSS-SECTION DETECTOR

It is recommended that carrier gases of small molecular size be used, such as H_2 or helium + 3% methane. The methane is necessary to quench the metastable helium which would yield anomalous results. Higher molecular weight gases would give less sensitivity.

CROSS-SECTION DETECTOR

Minimum Det. Quantity	-	20 ug
Response	-	Sensitive to all compounds (except carrier gas)
Linearity	-	5 x 10^5
Stability	-	Good (insensitive to temperature change)
Temperature Limit	-	225°C
Carrier Gas	-	H_2 or He + 3% CH_4

Summary: Simple, rugged. Less sensitive than T.C. detector. Employs radioactive source; limited to 225°C.

E. GAS DENSITY BALANCE

The gas density balance (G.D.B.) invented by Martin[14] was one of the first chromatographic detectors. The difficulties encountered in construction of this complicated gas density detector prevented its widespread use. In 1960, Nerheim[15] described a gas density balance which differed little in principle but was of greatly simplified construction. This model is now commercially available.

Gas density detectors have several advantages when used in gas chromatography:

1. Speed and simplicity
2. Calibration not required for quantitative analysis
3. Readily available carrier gases--N_2, A, CO_2
4. No sample destruction

The G.D.B. will operate in any chromatograph if the proper power supply is available. It is easiest to install in a dual column instrument since two regulated carrier flows are available.

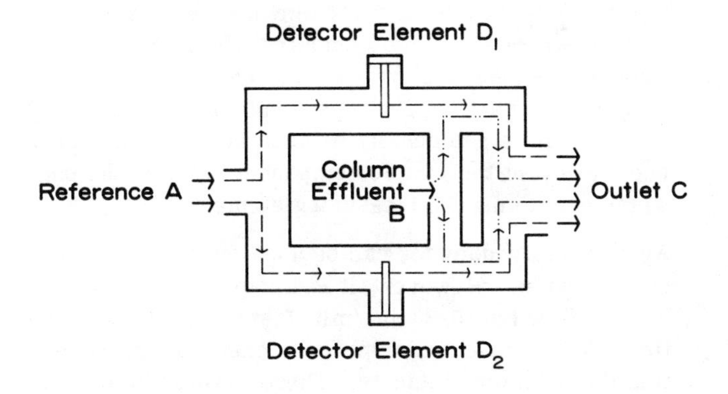

FIGURE V-16—GAS DENSITY BALANCE

Reference gas enters at A while gas from the column enters at B; a common exit is at C. Measuring elements are mounted in the reference stream at D_1 and D_2 and connected in a Wheatstone bridge. The sample never contacts the measuring elements which makes a G.D.B. useful for the analysis of corrosive materials. If the gas eluted at B has the same density as the reference gas, the gas flows are at equilibrium and no unbalance is detected by the bridge.

If the gas entering at B carries a sample component of higher density than the reference gas, the downward flow from B retards the lower gas flow AD_2 while the upper flow increases. This flow imbalance causes a variation in resistance of the measuring elements D_1 and D_2 causing an unbalance in the bridge. Solutes of lower density than the carrier gas tend to rise, producing the inverse phenomenon.

Sensitivity depends, in part, on the difference in density between the carrier gas and the sample component. Nitrogen is the preferred carrier gas except when CO, C_2H_4, C_2H_2 are being determined. Then carbon dioxide or argon should be used as carrier. Sulfur hexafluoride has been preferred by Guillemin and Auricourt as carrier gas [12]. Hydrogen and helium should not be used with a gas density balance since diffusion can occur allowing the sample component to enter the passages containing the measuring elements.

A gas density detector can be used over a wide range of flow rates. However, it is important that the reference flow be 15-20 ml/min faster than the column flow. This prevents sample components from contacting the detector elements. There is an optimum reference flow rate for maximum response. Column flow rate should be selected for the sample and column in use.

A variety of measuring elements can be used in a gas density detector--two or four W-2 filaments, two thermistors, or two "FilThe" elements (a combination of a thermistor and a filament). Density detectors utilizing the "FilThe" elements show greater sensitivity than those equipped with filaments or thermistors alone. Sensitivity of detectors with "FilThe" elements is comparable to a TC cell using thermistors. Sensitivity using filaments decreases linearly with temperature. Thermistors and "FilThe" elements show a logarithmic decrease in sensitivity with temperature.

D.M. Rosie has recommended that the following expression be used for measuring sensitivities of density detectors rather than the expression used for thermal conductivity detectors, S:

$$S_d = \frac{A \times B \times C \times D}{E \times F}$$

where: A = Area of peak in cm2
B = Flow rate of measuring carrier in ml/min
C = Reciprocal chart speed, min/cm
D = Recorder sensitivity, mv/cm
E = Vapor volume of sample, ml
F = Difference in molecular weight between sample and carrier gas.

A good approximation is that one "S_d" equals approximately fifteen "S". Thus, an "S_d" of 20 for a density detector would compare to a thermal conductivity detector with an "S" of 300.

The outstanding feature of a gas density detector is that quantitative analyses may be made without calibration. The gas density balance gives responses directly in weight percent, if peak areas are multiplied by a factor derived from the molecular weights of the solute and the carrier gas.

$$\text{Wt. of a component} = X \bullet A \bullet K$$

$$K = \frac{M_s}{M_s - M_{cg}}$$

where: A = Peak area
M_s = Molecular weight of the solute
M_{cg} = Molecular weight of the carrier gas
x = Constant for a given instrument

However, peaks must be completely resolved for the above to hold true, and, of course, the identity of the peak must be known. Gas density detectors can also be used for the determination of molecular weight. Phillips and Timms[16] were able to determine molecular weights of a variety of organic compounds with errors of $< \pm .8\%$.

GAS DENSITY BALANCE	
Minimum Detectable Quantity	Variable--dependent on measuring elements and carrier gas.
Response	All components whose molecular weight differs from carrier gas.
Linearity	10^3
Stability	Good
Carrier Gas	N_2, CO_2, A or others, except He or H_2
Temperature	Better sensitivity at temperatures less than 150°C
Summary:	Nondestructive, stable, moderate sensitivity, simple to operate. Requires good temperature and flow control.

BIBLIOGRAPHY

1. *Dimbat, M., Porter, PE., and Stross, F.H., Analytical Chemistry 28, 290 (1956).*
2. *Sternberg, J.C., Gallaway, W.S., and Jones, D.T.C., Gas Chromatography. Third International Symposium, Instrument Society of America, Academic Press. p. 231-267, 1962*
3. *Hartmann, C.H. and Oaks, D.M., Research Notes, Winter Issue, 1965*
4. *Hartmann, C.H., and Dimick, Keene P., Pittsburgh Conference on Analytical Chemistry and Applied Spectroscopy, Pittsburgh, Pennsylvania. March 1965.*
5. *Giuffrida, Laura, Jounal of the Association of Official Agricultrual Chemists 47, No. 2 1964*
6. *Coahran, D.R., American Chemical Society Regional Meeting Corvallis, Oregon. June 1965*
7. *Lovelock, J.E., Shoemaker, G.R., and Zlatkis, Albert, Analytical Chemistry 35, No. 4, April 1963*
8. *Simmonds, P.G. and Lovelock, J.E., Analytical Chemistry 35 No. 10, Sept. 1963.*
9. *Lovelock, J.E., Shoemaker, G.R., and Zlatkis, Albert, Analytical Chemistry 36, No. 8, July 1964*
10. *Claesson, S. Ark. Kemi. Min. Geol. A 23, No. 1 133 (1946)*
11. *Guillemin, C.L., Auricourt, F., J. Gas Chromatog. 1, No. 10:24 (1963)*
12. *Guillemin, C.L., Auricourt, F., J. Gas Chromatog. 2, 156 (1964)*
13. *Guillemin, C.L., Auricourt, R., J. Gas Chromatog. 4, 338 (1966)*
14. *Martin, A.J.P., James, A.T., Biochem. J. (London) 63, 138 (1956)*
15. *Nerheim, A.G., Patent No. 3,050,984, Standard Oil Co.*
16. *Phillips, C.S.G., Timms, P.L., J. Chromatog. 5, 131 (1961).*
17. *Dal Nogare, S., Juvet, R.S., Gas-Liquid Chromatography, 1962 page 205-210. Interscience, New York (1962).*
18. *Condon, R.D. et al, Gas Chromatograph 1960 p. 30, R.P. W. Scott, ed., Butterworths, Washington D.C. (1960).*

VI. QUALITATIVE ANALYSIS

A. INTRODUCTION

Gas chromatography is a separation technique. It has fantastic ability to separate components. Fortunately, the technique also allows identification, quantitation, and collection of these compounds.

One of the problems currently facing chromatographic workers is the positive identification of the numerous peaks emerging from G.C. columns. The use of capillary and high efficiency packed columns has succeeded in giving astonishing separations. Unfortunately, our ability to identify each compound has not progressed so rapidly. One can employ both chromatographic and non-chromatographic techniques; chromatographic techniques use data available from gas chromatographs and accessories.

B. CHROMATOGRAPHIC IDENTIFICATION

1. Retention data

The volume of carrier gas required to elute a compound from the G.C. column is called the retention volume.

Under constant pressure conditions, the flow rate is linear with time and one could also speak of retention time. This retention volume or time is characteristic of the sample and the liquid phase, and can therefore be used to identify the sample. Column temperature, however, must remain constant. Identification is based on a comparison of the retention time of the unknown component with that obtained from a known compound analyzed under identical conditions.

With well designed instruments the retention time is quite reproducible. See Table VI-1.

TABLE VI-1—RETENTION TIME REPRODUCIBILITY

Compound	Retention time, seconds					
	Run I	Run II	Run III	Run IV	Run V	Ave.
C7	231	231	232	230	231	231
C8	302	301	305	300	302	302
C9	389	389	395	387	390	390
C10	521	522	528	519	523	523
C12	863	864	868	864	865	865
C14	1191	1190	1195	1193	1192	1192

Figure VI-1 shows chromatograms of an unknown alcohol mixture and a solution of alcohol standards analyzed under the same conditions. By comparing the chromatograms it is possible to identify peaks 2, 3, 4, 7 and 9 as methyl, ethyl, n-propyl, n-butyl and n-amyl alcohols. It is possible for different compounds to have identical, or very close retention times. In these cases, confirmatory identification by infrared, mass spectroscopy or nuclear magnetic resonance is recommended.

a. Uncorrected retention volume is that volume measured from point of injection to the peak maximum. This is not commonly used since data cannot be compared with other columns or instruments. Uncorrected retention volume depends upon:

(1) Column dimensions (length and diameter)
(2) Liquid phase (type, amount)
(3) Column temperature
(4) Flow rate
(5) Type of carrier gas
(6) Instrument dead volume
(7) Pressure drop

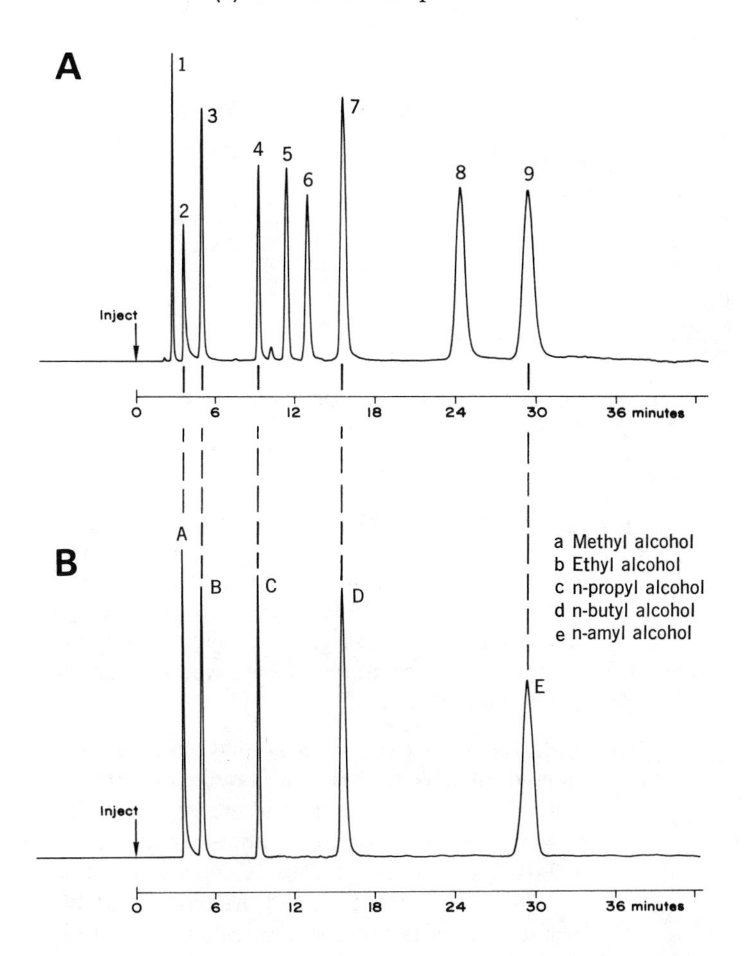

FIGURE VI-1—IDENTIFICATION OF UNKNOWN PEAKS BY USE OF STANDARDS

b. Adjusted retention volume is that volume measured from the air peak (leading edge of solvent peak for ionization detectors) to the peak maximum. This parameter depends upon the same variables listed above except that it is corrected for instrument dead volume.

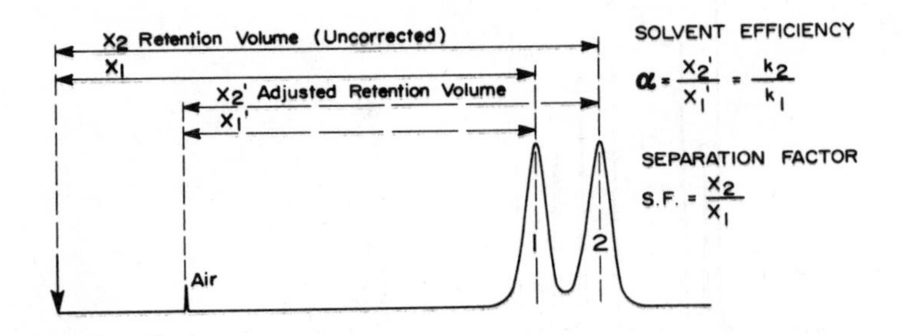

FIGURE VI-2—RETENTION VOLUME CALCULATIONS

c. Corrected retention volume is that volume corrected for the pressure drop across the column. This allows for the fact that gases are compressible. It is also not commonly used since it depends upon accurate control of column temperature, flow rate, and liquid phase conditions.

d. Relative retention, α , is that volume reported relative to that of a standard component. It is more reliable and easier to obtain than corrected retention volume. In general, relative retention volume is dependent only on column temperature and the type of liquid phase. This is the recommended method to identify peaks.

2. Identification by log plotting of homologous series

a. Semi-log plotting, one column.

If a sample containing several members of a homologous series is injected into a gas chromatograph, a plot of the log of the retention times is proportional to some increasing property of the homologous series. Therefore, identification of members of a homologous series can be obtained by plotting log of the retention time vs. the number of carbon atoms, number of methylene groups, boiling point, etc.

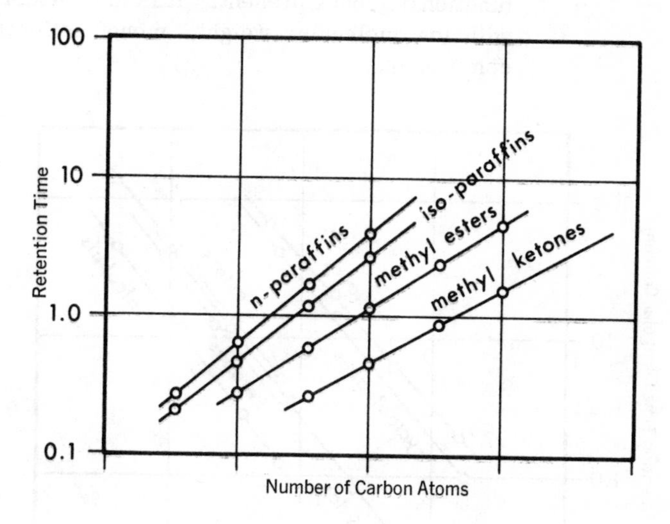

FIGURE VI-3—LOG RETENTION TIME VS. CARBON NUMBER

This method of identification is advantageous in that only 2 or 3 compounds are needed to establish the slope of the line and thus can be used to identify other members of the same series.

b. Log-log plotting, two columns.

When a plot is made of retention time on one liquid phase against retention time on another liquid phase of different polarity, straight lines are formed for each homologous series, the slopes of which are characteristic of the functional groups.

By chromatographing a substance with two different columns, a substance can be characterized, though not necessarily identified, by a point in an area rather than a point on a straight line. This is useful in indicating the functional groups present. It is most useful with low molecular weight, monofunctional compounds.

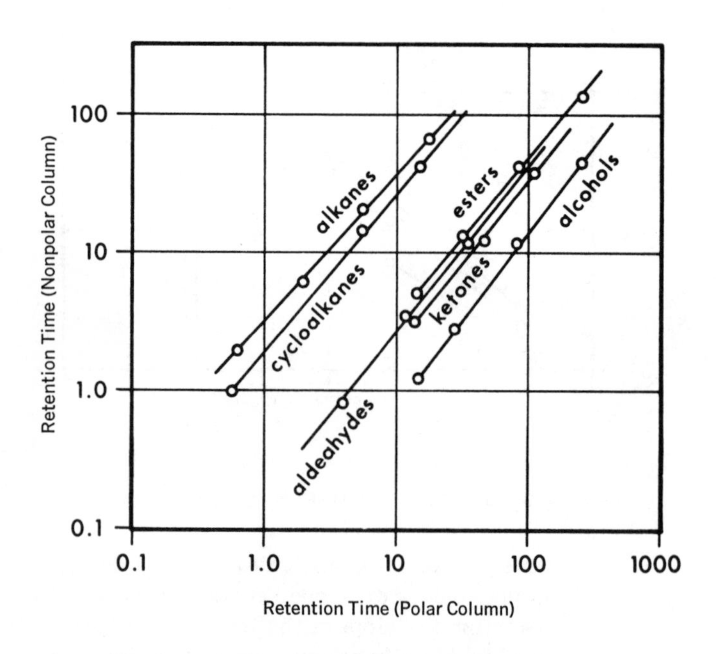

FIGURE VI-4—LOG RETENTION TIME ON TWO COLUMNS

3. Retention index, I

Peak identification can be confirmed by the use of the retention index, I, of Wehrli and Kovats [5].

$$I = 200 \frac{\log \alpha(x)}{\log \alpha_{(P_z + 2)}} + 100\,z$$

where: α_x is the relative retention of substance x referred to a normal paraffin, Pz, containing an even number of carbon atoms. The quantity, $\alpha_{(Pz+2)}$ is the relative retention of a normal paraffin,(Pz + 2) referred to the normal paraffin, Pz. The number of carbon atoms is denoted by z.

The retention indices of the even-numbered paraffins are defined as 100z for every temperature and for every liquid phase.

Information on the structure of an unknown peak can be obtained from:

$$\Delta I = I_{polar} - I_{nonpolar}$$

where: Δ I is defined as the difference between the retention index of a compound on a polar liquid phase, I_{polar}, and the retention index of a compound on a nonpolar phase. Δ I values have been measured for a vast number of aliphatic, alicyclic and aromatic compounds by Wehrli and Kovats.

4. Identification by relative detector response (Dual Channel)

The comparison of response ratios of a given compound analyzed by two different detectors under fixed conditions is characteristic of that compound. One sample is chromatogrammed on one column, and the column effluent split to feed two different detectors. Frequently used combinations are:

a. Flame ionization and electron capture
b. Flame ionization and radioactivity detector
c. Flame ionization and phosphorus detector
d. Phosphorus and electron capture

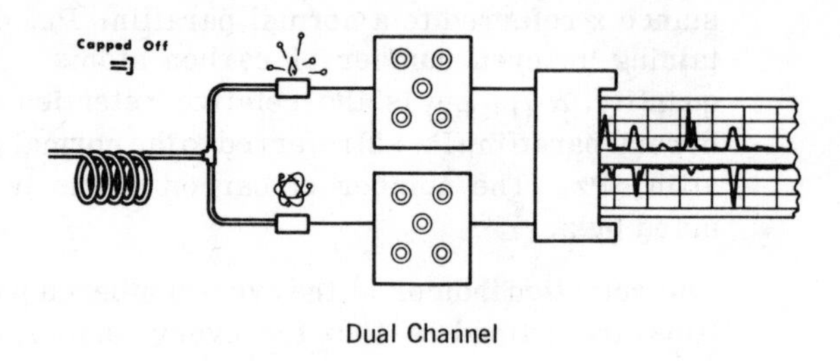

Dual Channel

FIGURE VI-5—SCHEMATIC OF DUAL CHANNEL SYSTEM

C. NON-CHROMATOGRAPHIC IDENTIFICATION

1. Classical microchemical tests can also be used to identify chromatography peaks. The column effluent can be bubbled into solutions, and sensitive color reactions can help identify the type of compound. This information, together with molecular weight, or boiling point from a log retention time plot is often sufficient to identify the specific compound.

2. <u>Identification by derivative formation</u>
The determination of melting points of derivatives is a useful qualitative technique. Three books with melting point tables recommended are [2,3, 4].

TABLE VI-2—IDENTIFICATION BY FUNCTIONAL GROUP CLASSIFICATION TESTS[1]

COMPOUND	REAGENT	REACTION	DETECTION LIMIT (μg)
Alcohols	Ceric Nitrate	amber color	100
Aldehydes	2, 4 DNP	yellow ppt.	20
	Schiff's	pink color	50
Ketones	2, 4 DNP	yellow ppt.	20
Esters	Ferric Hydroxamate	red color	40
Mercaptans	Sodium Nitroprusside	red color	50
Sulfides	Sodium Nitroprusside	red color	50
Disulfides	Sodium Nitroprusside	red color	50
Amines	Sodium Nitroprusside	red color, 1° blue color, 2°	50
Nitriles	Ferric Hydroxamate-Propylene Glycol	red color	40
Aromatics	HCHO - H_2SO_4	red-wine color	20
Aliphatic unsat.	HCHO - H_2SO_4	red-wine color	20
Alkyl Halide	Alc. $AgNO_3$	white ppt.	20

3. Identification by auxiliary instrumentation

a. Infrared

This is an extremely useful method of positive identification. Samples may be passed directly into gas or liquid sampling cells for easy handling. The limit of detection is about 0.5 microliter of acetone without scale expansion. To calculate limits for other components, compare their molar absorptivities with acetone. In its present state IR analysis cannot be used with capillary columns.

b. Mass spectrometry

Once the component has been separated and trapped in sufficient quantity to enable easy handling, mass spectrometry is a very useful tool. This technique provides positive identification except for isomers. It does, however, require the use of an expensive instrument which many laboratories cannot afford. Temperature limits are presently set by the ability to avoid condensation in connecting lines and memory effects in the sample inlet system of the M.S. Work, however, has been done up to Mass 1800. Sensitivity is sufficient to enable analysis of peaks in the low microgram range.

c. Nuclear magnetic resonance
With increasing sensitivity, the sample size requirement decreases rapidly. This method is becoming more popular in identifying G.C. peaks. Present sample requirements are in the 1-10 mg range, and identification is positive, particularly for structural isomers.

4. Coulometry, polarography, flame photometry, ultraviolet, visible spectroscopy, and elemental analysis are also some of the methods available for identification.

BIBLIOGRAPHY

1. *Walsh, J.T. and Merritt, C., Jr., Analytical Chemistry 32, 1378 (1960.*
2. *Shriner, R.L., Fuson, R.C., and Curtin, D.Y., Systematic Identification of Organic Compounds, Fourth Edition, Wiley, 1956.*
3. *Cheronis, N.D. and Entrikin, J.B., Semimicro Qualitative Organic Analysis, Third Edition, Wiley (Interscience) 1964.*
4. *Technique of Organic Chemistry, Volume XI, Part I, Elucidation of Structure by Physical and Chemical Methods, Interscience Publishers, 1963.*
5. *Wehrli, A., and E. sz. Kovats, Helv. ChimActa, 42, 2709 (1959)*

VII. QUANTITATIVE ANALYSIS

It is estimated that there are more than 50,000 gas chromatographs in the world. The primary application of these instruments is the quantitative analysis of gases, liquids, and solids. Extremely high accuracy can be obtained provided the proper techniques are employed. This chapter will discuss possible sources of error, calculations, integration techniques, and statistical treatment of data.

A. POSSIBLE SOURCES OF ERROR

Taken in sequence the possible areas where errors can be introduced in the chromatographic technique are: (1) Sampling technique, (2) Sample adsorption or decomposition in the chromatograph, (3) Detector performance, (4) Recorder performance, (5) Integration technique, and (6) Calculations.

1. Sampling technique

   There are two sources of error in the sampling technique. The first problem is to take exactly that sample which you wish to analyze. Consider a sample of crude oil - do you wish to sample the gas phase or the liquid phase, or any solids, should they be present? This is a classical sampling problem and can be a source of error in many G. C. methods. The second problem in the sampling technique is to make sure that the sample you have taken actually gets into the gas chromatograph. Does the sample decompose or vaporize, or undergo some reaction from the time that you sample until it is placed into the gas chromatograph? These points may seem obvious, but many workers ignore these items and their quantitative results suffer.

2. Sample adsorption or decomposition

The second source of error lies in the assumption that all of the sample injected actually produces the peaks which you observe. It can happen that compounds are decomposed or adsorbed in the injection port, on the column, or in the detector. There are cases in the literature where an entire sample is irreversibly adsorbed in the chromatographic system. Quantitative G. C. demands that all of the sample which you injected produced the peaks which you integrate. Reproducible loss may allow calibration curves to be constructed and thus compensate for this source of error. One check is the blending of a difficult compound with an inert hydrocarbon and injecting different dilutions of this mixture. The area ratios should remain constant.

3. Detector performance

Each detector responds differently to different compounds. These response factors must be known. In addition, as operating conditions change, the detector response also changes. This influence must be known and new calibration curves constructed. In the T. C. cell, for example, the response is due to differences in thermal conductivity between pure carrier gas, and a mixture of carrier gas and the sample. For accurate and reproducible analysis, the purity of carrier gas, gas flow rate, detector temperature, filament current, filament resistance, and pressure inside the detector must remain constant. If one of these conditions changes drastically, so will the detector performance.

4. Recorder performance

A recorder is a mechanical electrical device, and like any other instrument is subject to error. For accurate quantitative results, one must check the linearity range, pen speed, dead band, and electrical zero. All of these factors can affect the quantitative results. Where the chromatogram is used for quantitative measurements, the recorder is a possible source of error and the precision obtainable with standards should be determined.

5. Integration technique

Probably the most critical step is the conversion of the chromatographic peak into numbers related to the sample composition. This step is the conversion of the analog peak into digital form. This problem is discussed in detail in section C.

6. Calculations

Once the numbers representing the area are obtained, they must be related to the composition of the sample. This is discussed separately in the next section.

B. CALCULATIONS

1. Area normalization

By normalizing, we mean calculating the percent composition by measuring the area of each peak and dividing the individual areas by the total area, e.g.,

$$\% A = \frac{\text{Area of A}}{\text{Total area}} \bullet 100$$

When analyzing close boiling components of a homologous series, this method can be used to calculate weight percent. This assumes that all peaks were eluted, and that each compound has the same detector response. If these assumptions have been tested, this method is rapid and simple.

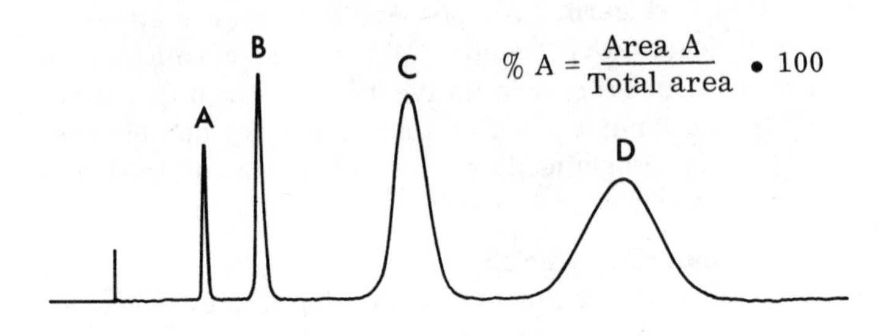

FIGURE VII-1—AREA NORMALIZATION

2. Correction factors

Areas of compounds are not directly proportional to the percent composition, i.e., different compounds have different detector responses; therefore, it is necessary to determine correction factors. Once determined, these correction factors can be used to calculate the percent composition. Since detectors operate on different principles, different factors must be calculated for different detectors. Depending upon the calibration method employed, the correction factor can be used to obtain the weight percent, volume percent, or mole percent of the components desired.

a. A method for calculating FID response factors is as follows:

A standard solution of compounds a, b, c, d, and e is prepared and produces the chromatogram shown in Figure VII-2. The weights "W" injected are known. The areas "A" are

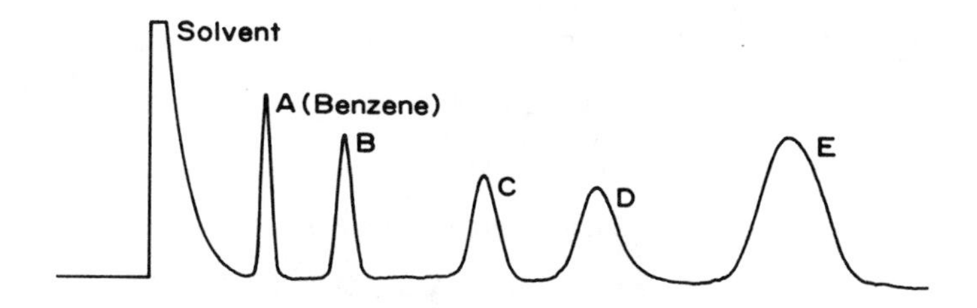

FIGURE VII-2—CALCULATION OF FID RESPONSE FACTORS

measured. The ratio A/W is calculated for each peak. The correction factor "F" is calculated by dividing the A/W of each peak by the benzene A/W. These factors are relative to benzene, i.e., the benzene factor is arbitrarily set equal to 1.00.

TABLE VII-1—PROCEDURE FOR CALCULATION OF FID FACTORS

Peaks	W Wt. Injected μg	A Area cm^2	A/W	F Correction Factors
a (benzene)	.435	4.0	9.19	1.00
b	.653	6.5	9.95	1.08
c	.864	7.6	8.79	.96
d	.864	8.1	9.38	1.02
e	1.760	15.0	8.52	.93

Under the same detector conditions, these factors can be used time and time again to calculate the weight percent of "b", "c", "d", and "e" relative to "a" (benzene).

From these results the weight of an unknown "b" can be calculated:

$$W_b = \frac{W_a \bullet A_b}{F_b \bullet A_a}$$

where:

W_b = weight of component b

W_a = weight of standard a

A_a = measured area of standard a

A_b = measured area of component b

F_b = correction factor of compound "b" relative to compound "a" at equal weights.

The response of a FID is independent of temperature, carrier gas, and flow rate. This makes it well suited, possibly the best detector, for quantitative analysis. For weight percent calculations, Table VII-2 can be used assuming the factor for benzene is 1.00[1,2,3].

An excellent reference for response factors for both flame ionization and thermal conductivity detectors is the article by Dietz[4].

TABLE VII-2—FID RESPONSE FACTORS [4]

COMPOUND	WEIGHT FACTOR	COMPOUND	WEIGHT FACTOR
<u>n-Alkanes</u>		<u>iso-Alkanes</u>	
Methane	1.23	C_4	1.11
Ethane	1.15	C_5	1.11
Propane	1.13	C_6	1.10
Butane	1.11	C_7	1.10
Pentane	1.11	C_8	1.10
Hexane	1.11	C_9	1.09
Heptane	1.10	C_{10}	1.09
Octane	1.10	<u>Monoalkenes</u>	
Nonane	1.09	C_2	1.08
		C_3	1.08
Decane	1.09	C_{10}	1.08

TABLE VII-2—FID RESPONSE FACTORS (cont.)

COMPOUND	WEIGHT FACTOR	COMPOUND	WEIGHT FACTOR
Dialkenes		Esters, n-, iso-	
C_3	1.02	C_2	2.30
C_4	1.04	C_3	1.90
C_5	1.05	C_4	1.69
C_6	1.05	C_5	1.57
C_7	1.05	C_6	1.48
C_8	1.06	C_7	1.43
C_9	1.06	C_8	1.39
C_{10}	1.06	C_9	1.35
		C_{10}	1.32
Naphthenes			
Cyclopentane	1.08		
Methylcylopentane	1.08		
Cyclohexane	1.08	Ethers, n-, iso-unsymmetrical, symmetrical	
Methylcyclohexane	1.08		
All isomers	1.08	C_2	1.77
Aromatics		C_3	1.54
		C_4	1.42
Benezene	1.00	C_5	1.35
Toluene	1.01	C_6	1.31
Ethylbenzene	1.02	C_7	1.27
Xylenes	1.02	C_8	1.25
C_3 aromatics	1.03	C_9	1.23
C_4 aromatics	1.03	C_{10}	1.21
C_5 aromatics	1.03		
Alcohols, primary, secondary, tertiary		Ketones, n-, iso-unsymmetrical symmetrical	
Methanol	2.46		
Ethanol	1.77	C_3	1.48
C_3 alcohols	1.55	C_4	1.38
C_4 alcohols	1.42	C_5	1.32
C_5 alcohols	1.35	C_6	1.28
C_6 alcohols	1.30	C_7	1.25
C_7 alcohols	1.27	C_8	1.23
C_8 alcohols	1.25	C_9	1.21
C_9 alcohols	1.23	C_{10}	1.20

Under normal conditions, the FID does not respond to H_2O, H_2S, CS_2, COS, HCN, HCOOH, $(COOH)_2$, NH_3, NO, N_2O, NO_2, CO, CO_2, O_2, N_2, rare gases, halogens, hydrogen halides, or interhalogen compounds.

b. The TC factors can be calculated in the same manner that the FID factors were. To use the weight factors, the area of the peak is measured and multiplied by the weight factor to give the true weight area. These values are normalized and the results are weight percent of each compound. The weight factor is determined by dividing the molecular weight by the relative thermal response per mole. This is calculated relative to benzene whose thermal response is 100[4,5].

Example: A mixture of ethanol, heptane, benzene, and ethyl acetate were analyzed using a thermal conductivity detector. The problem is to determine the weight percent of each component if their respective peak areas were 5.0, 9.0, 4.0, and 7.0 cm^2.

Calculation of weight % for T. C. Detector.

Step #1

COMPOUND	AREA (cm^2)	WEIGHT FACTOR
Ethanol	5.0	0.64
Heptane	9.0	0.70
Benzene	4.0	0.78
Ethyl Acetate	7.0	0.79

Multiply the peak area by the weight factor.

Step #2

Ethanol (5.0)(0.64)	= 0.320
Heptane (9.0)(0.70)	= 0.630
Benzene (4.0)(0.78)	= 0.312
Ethyl Acetate (7.0)(0.79)	= 0.553
TOTAL	1.815

Now normalize and the answer is weight %.

Step #3

NORMALIZE		WEIGHT %
$\frac{0.320}{1.815}$	=	17.6
$\frac{0.630}{1.815}$	=	34.7
$\frac{0.312}{1.815}$	=	17.2
$\frac{0.553}{1.815}$	=	30.5
		100.0

TABLE VII-3—T.C. WEIGHT FACTORS[4]

COMPOUND	WEIGHT FACTOR	COMPOUND	WEIGHT FACTOR
Normal Paraffins		Unsaturates	
Methane	0.45	Ethylene	0.585
Ethane	0.59	Propylene	0.652
Propane	0.68	Isobutylene	0.683
Butane	0.68	Butene-1	0.697
Pentane	0.69	trans-Butene-2	0.658
Hexane	0.70	cis-Butene-2	0.643
Heptane	0.70	3-Methylbutene-1	0.707
Octane	0.71	2-Methylbutene-1	0.707
Nonane	0.72	Pentene-1	0.710
Decane	0.71	trans-Pentene-2	0.673
Undecane	0.79	cis-Pentene-2	0.710
Tetradecane	0.85	2-Methylpentene-2	0.729
C_{20} to C_{36}	0.72	2,4,4-Trimethyl-pentene-1	0.71
		Propadiene	0.76
		1,3-Butadiene	0.674
Branched Paraffins		Cyclopentadiene	0.97
Isobutane	0.710	Isoprene	0.738
Isopentane	0.707	1-Methylcyclohex-ene	0.837
Neopentane	0.727		
2,2-Dimethylbutane	0.741	Methylacetylene	0.69
2,3-Dimethylbutane	0.741	Dicyclopentadiene	1.73
2-Methylpentane	0.714	4-Vinylcyclohexene	0.83
3-Methylpentane	0.725	Cyclopentene	0.844
2,2-Dimethylketone	0.752		
2,4-Dimethylpentane	0.775	Aromatics	
2,3-Dimethylpentane	0.741	Benzene	0.780
3,5-Dimethylpentane	0.750	Toluene	0.794
2,2,3-Trimethylbu-tane	0.775	Ethylbenzene	0.822
2-Methylhexane	0.735	meta-Xylene	0.812
3-Methylhexane	0.752	para-Xylene	0.812
3-Ethylpentane	0.763	ortho-Xylene	0.840
2,2,4-Trimethyl-pentane	0.775	Isopropylbenzene	0.847
		n-Propylbenzene	0.826
		1,2,4-Trimethyl-benzene	0.800

TABLE VII-3—T.C. WEIGHT FACTORS(4) (cont.)

COMPOUND	WEIGHT FACTOR	COMPOUND	WEIGHT FACTOR
Aromatics (cont.)		Hetero Compounds (cont.)	
1,2,3-Trimethyl-benzene	0.806	Propylene oxide	0.730
p-Ethyltoluene	0.800	Hydrogen sulfide	0.890
1,3,5-Trimethyl-benzene	0.805	Methyl mercaptan	0.810
sec.-Butylbenzene	0.847	Ethyl mercaptan	0.713
Biphenyl	0.912	1-Propanethiol	0.750
1,2-Diphenylbenzene	1.060	Tetrahydrofuran	0.870
1,3-Diphenylbenzene	1.00	Thiophane (cyclic sulfide)	0.855
1,4-Diphenylbenzene	1.03	Ethyl silicate	0.995
Triphenylmethane	1.05	Acetaldehyde	0.680
Naphthalene	0.923	Cellosolve	0.840
Tetralin	0.910		
1-Methyltetralin	0.927		
1-Ethyltetralin	0.944		
trans-Decalin	0.920		
cis-Decalin	0.913	Nitrogen Compounds	
		n-Butylamine	0.64
		n-Pentylamine	0.57
		Pyrrole	0.78
Inorganic Compounds		Pyrroline	0.83
Argon	0.95	Pyrrolidine	0.78
Nitrogen	0.67	Pyridine	0.79
Oxygen	0.80	1,2,5,6-Tetra-hydropyridine	0.81
Carbon dioxide	0.915	Piperidine	0.83
Carbon monoxide	0.67	Acrylonitrile	0.68
Carbon tetrachloride	1.43	Propionitrile	0.65
Iron carbonyl ($Fe(CO_5)$)	1.30	n-Butyronitrile	0.66
Hydrogen sulfide	0.89	Aniline	0.82
Water	0.55	Quinoline	0.67
		trans-Decahydro-quinoline	1.19
Hetero Compounds		cis-Decahydro-quinoline	1.19
Pyrrole	0.780		
Hexylamine	0.970		
Ethylene oxide	0.758	Ammonia	0.42

TABLE VII-3—T.C. WEIGHT FACTORS[4] (cont.)

COMPOUND	WEIGHT FACTOR	COMPOUND	WEIGHT FACTOR
Oxygenated Compounds		Oxygenated Compounds (cont.)	
Ketones		Alcohols (cont.)	
Acetone	0.68	n-Hexanol	0.87
Methyl ethyl ketone	0.74	3-Hexanol	0.80
Diethyl ketone (3-pentanone)	0.78	2-Hexanol	0.77
3-Hexanone	0.81	n-Heptanol	0.91
2-Hexanone	0.77	Decanol-5	0.86
3, 3-Dimethylbutanone-2	0.97	Dodecanol-2	0.93
Methyl-n-amylketone	0.86	Cyclopentanol	0.79
Methyl-n-hexylketone	0.87	Cyclohexanol	0.89
Cyclopentanone	0.79	Acetates	
Cyclohexanone	0.785	Ethyl acetate	0.79
2-Nonanone	0.84	Isopropyl acetate	0.84
Methyl isobutyl ketone	0.86	n-Butyl acetate	0.86
Methyl isoamyl ketone	0.83	n-Amyl acetate	0.89
Alcohols		iso-Amyl acetate	0.90
Methanol	0.58	n-Heptyl acetate	0.93
Ethanol	0.64	Ethers	
n-Propanol	0.72	Diethyl ether	0.67
Isopropanol	0.71	Diisopropyl ether	0.79
n-Butanol	0.78	Di-n-propyl ether	0.78
Isobutanol	0.77	Ethyl-n-butyl ether	0.79
sec-Butanol	0.76	Di-n-butyl ether	0.81
tert-Butanol	0.77	Di-n-amyl ether	0.86
3-Methylpentanol-1	0.80	Diols	
2-Pentanol	0.80	2, 5-Hexanediol	0.93
3-Pentanol	0.81	1, 6-Hexanediol	0.98
2-Methyl-2-butanol	0.83	1, 10-Decanediol	1.62
		1, 12-Decanediol	1.58
		C_{14} Diol (from sec. C_{14} alcohol)	1.80

3. Absolute calibration

Known amounts of the compounds in question are prepared and chromatographed. The value of peak areas are then plotted against the weight of sample injected. An exact weight of the unknown sample is then injected and its peak area is compared to that of the standard. The results can be interpreted graphically or calculated by the formula:

$\% B = (Area_B)(K)$, where K is the ratio of

$$\frac{\% \text{ of standard B}}{Area_B}$$

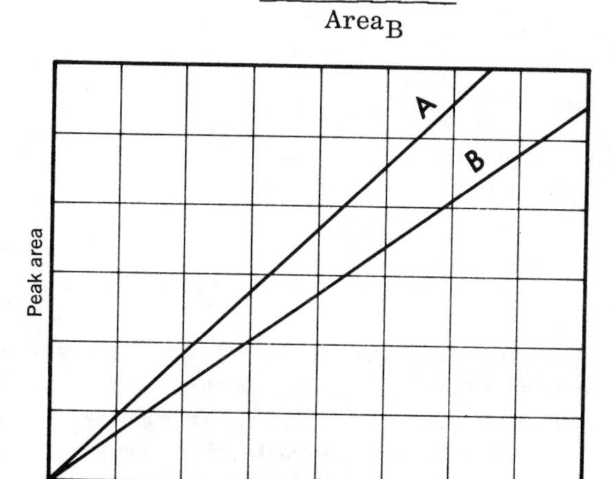

FIGURE VII-3—ABSOLUTE CALIBRATION

Peak heights, in addition to peak areas, may also be employed. Disadvantages of direct calibration are that the precise amount of sample injected must be known and that calibration is time consuming. Also, the sensitivity of the detector must remain constant from run to run and day to day in order to compare results with the calibration graph.

4. Internal standardization

This method is also known as relative or indirect calibration. Known weight ratios of the sample and a standard are prepared and chromatographed. The peak areas are measured. Area ratios are plotted against weight ratios to obtain a graph.

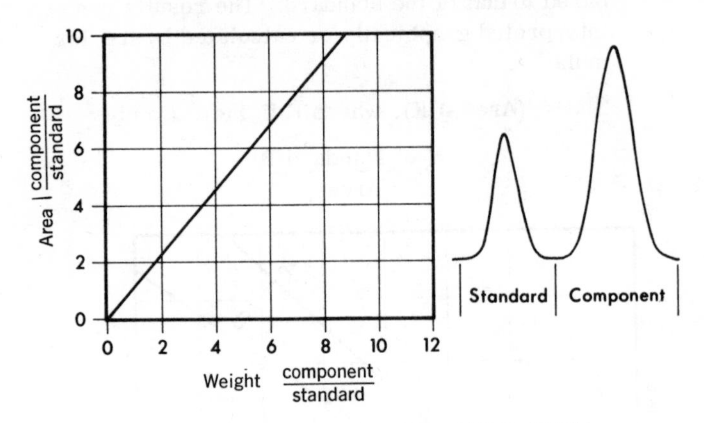

FIGURE VII-4—RELATIVE CALIBRATION CURVE

An accurately known amount of the internal standard is then added to the unknown sample and this mixture chromatographed. Area ratios are measured and from the calibration graph the weight ratio of the unknown to the standard is read. Since the amount of standard added is known, it is a simple calculation to determine the amount of the unknown compound present.

Example: To the unknown, add 5 ml of a solution containing the standard whose concentration is 100 μg/ml. Upon chromatographing the mixture, the area ratio is found to be 8; therefore, the weight ratio is 7 (see Figure VII-4). Knowing the standard concentration to be 100 μg/ml, then the component concentration is 7 x 100 μg/ml.

Since we added 5 ml of standard solution, then the total amount of the unknown is 5 ml x 700 $\frac{\mu g}{ml}$ = 3500 μg or 3.5 mg.

The advantages of this calibration method are that quantities injected need not be accurately measured and the detector response need not be known or remain constant since any change in response will not alter the area ratio. The chief disadvantage of this method is the difficulty in finding a standard that does not interfere with a component in the sample.

Requirements for an internal standard are:

a. Must be well resolved from other peaks.

b. Must elute close to peaks of interest.

c. Should be approximate concentration of unknown.

d. Structural similarity to unknown.

C. INTEGRATION

There are many ways for relating the peak shape to the sample concentration. Figure VII-5 shows the results of a survey of 1400 chemists in the United States. These results indicate that the most popular methods are in decreasing order: peak height, disc integrator, triangulation, planimetry, electronic digital integrator, and cutting and weighing paper. All of these methods will be discussed and compared.

Peak height measurement is more rapid than peak area; however, plots of peak height vs. sample size have a more limited linear range than corresponding plots for peak area. Peak heights and widths are frequently dependent on sample size and sample feed volume;

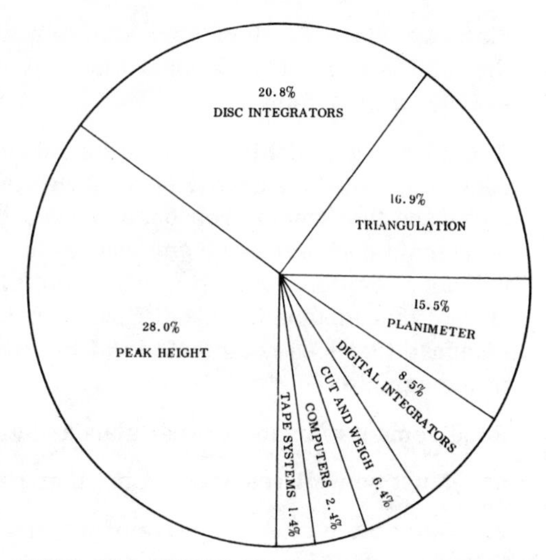

FIGURE VII-5—SURVEY OF QUANTITATIVE TECHNIQUES

however, peak area is not. Peak heights are generally used if samples are less than 10 μg for packed columns and 0.1 μg for capillary columns.

Peak height is measured, usually in mm, as the distance from the baseline to peak maxima as shown in Figure VII-6. If the baseline drifts, as in peak D, draw the best line between start and finish of the peak.

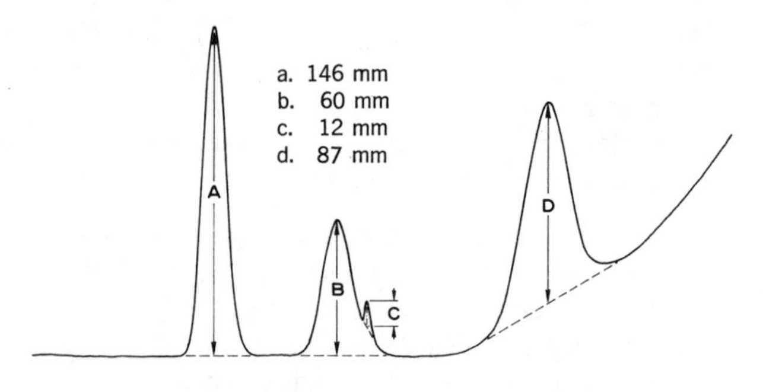

FIGURE VII-6—PEAK HEIGHT MEASUREMENT

Peak areas are less dependent than peak height on operating conditions. They are certainly the most widely used at the present, and the remainder of our discussion will concentrate on peak area measurement.

1. Planimetry
   The peak is traced manually with a planimeter. A planimeter is a mechanical device which measures area by tracing the perimeter of the peak. The area is presented digitally on a dial. This technique is tedious, time-consuming, and less precise than many other methods (5). The reproducibility between different persons is poor. Precision can be improved by tracing each peak several times and taking an average value.

2. Height x width at half-height
   Since normal peaks approximate a triangle, one could approximate the area by multiplying the peak height times the width at half-height. The normal peak base is not taken since large deviations may be observed due to tailing or adsorption. This technique is rapid and simple. The results are good with symmetrical peaks of reasonable width. A fast recorder chart speed should be used to make peak width measurements more accurate.

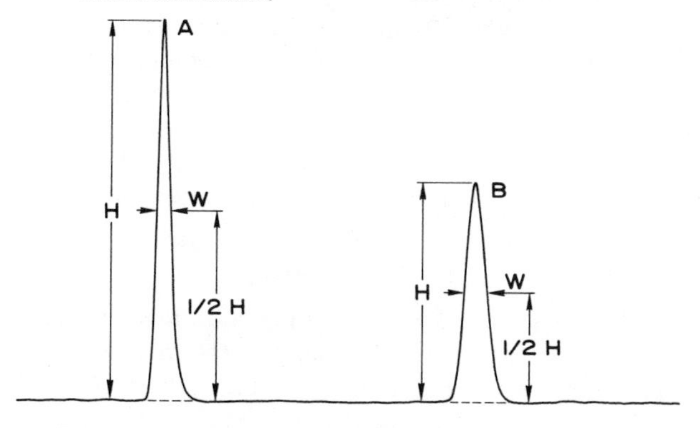

FIGURE VII-7—PEAK HEIGHT MEASUREMENT BY H x W

3. Triangulation
Height is measured from the baseline to the intersection of the two tangents (see Figure VII-8). The base is taken as the intersection of the two tangents with the baseline. The area is calculated by the triangle formula, area = 1/2 BH. This method is time consuming but the precision is reasonable provided the peak shapes allow significant measurements. The peak must also have a gaussian shape.

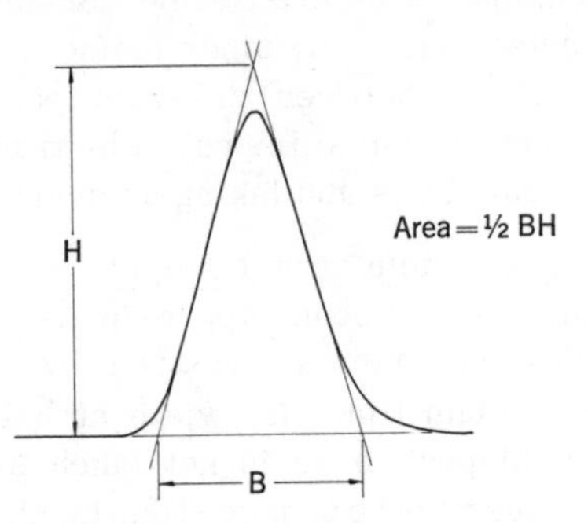

FIGURE VII-8—MEASUREMENT OF AREA BY TRIANGULATION

4. Cutting and weighing paper
Peak areas are determined by cutting out the chromatographic peak and weighing the paper on an analytical balance. This method is quite time consuming but can be fairly precise, particularly for asymmetrical peaks. A disadvantage is that the chromatographic data is destroyed. Care must be exercised in cutting the peaks and the thickness and moisture content of the chart paper must be constant.

5. Disc integrator
The simple Ball disc integrator manufactured by Disc Instrument Company is the most widely employed integrator in gas chromatography. This

system is illustrated schematically in Figure VII-9. The integrator pen is linked mechanically to the ball which rides on the rotating disc. When the recorder pen deflects, the ball moves away from the center of the disc and the ball begins to rotate. The rotation of the ball is transmitted mechanically to the integrator pen. Since the disc rotates at a rate proportional to the time base of the recorder chart, the integrator pen traces lines which represent the area of the recorder pen's travel. This method is both precise and rapid. It appears from our experience that the limit of precision of the disc integrator is the mechanical performance of the recorder itself. The reproducibility of data between different chemists and different laboratories with the disc integrator is quite good.

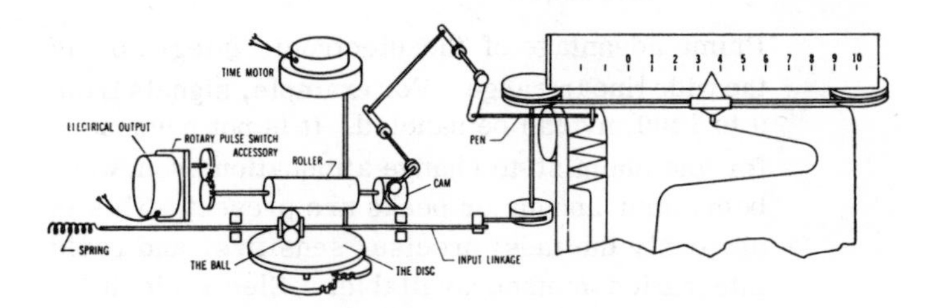

FIGURE VII-9—SCHEMATIC OF DISC INTEGRATOR

Procedure, together with examples for reading disc traces is found in the Appendix.

6. <u>Electronic digital integrator</u>
The chromatographic input signal is fed into a voltage to frequency converter which generates an output pulse rate proportional to the peak area. When the slope detector senses a peak, the pulses from the V - F converter are accumulated and printed out as a measure of the peak area. These

integrators have a wide linear range, a high count rate, and sensitive power detection. In the sophisticated units, a retention time is also recorded and printed out.

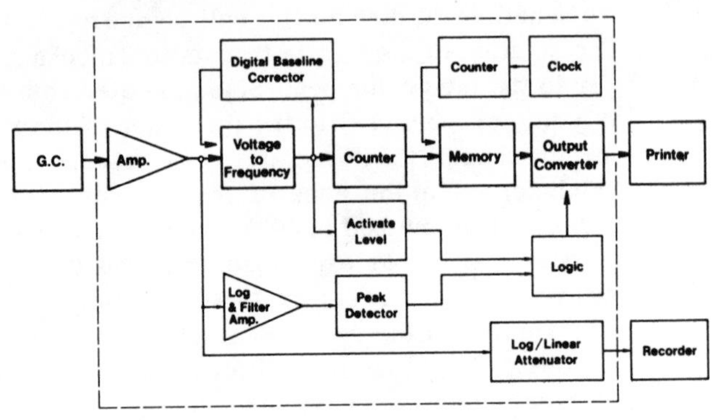

FIGURE VII-10—BLOCK DIAGRAM OF AN ELECTRONIC DIGITAL INTEGRATOR

Prime advantage of the electronic integrator is the wide linear range. For example, signals from 0 to 1400 mv can be handled. It is not necessary for the chemist to change attenuation even when both minor and major peaks are present. This is obviously the most precise, sensitive, and rapid integration method available. Electronic integrators are expensive ($3000 to $7000) but the cost can often be justified based upon the accuracy and increased data output.

Our research laboratory in Walnut Creek has compared different integration techniques. The sample chosen for the study was Phillips Hydrocarbon Mix #37 (see Figure VII-11). It seems a typical sample: it contains 8 components of quite different concentrations; some peaks are well resolved; however, butylene-1 and iso-butylene are not resolved; peak widths vary from 20 to 60 seconds at the base.

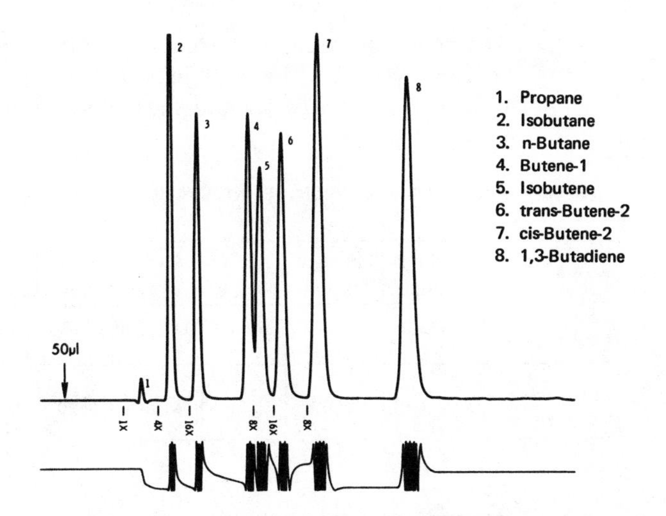

FIGURE VII-11—SAMPLE FOR INTEGRATION STUDY

The gas chromatographic signal was connected directly to the electronic integrator. The integrator attenuator was used to control the output signal to the recorder equipped with a Disc integrator. This system allows the recorder peaks, the disc trace, and the electronic integrator data to be made simultaneously.

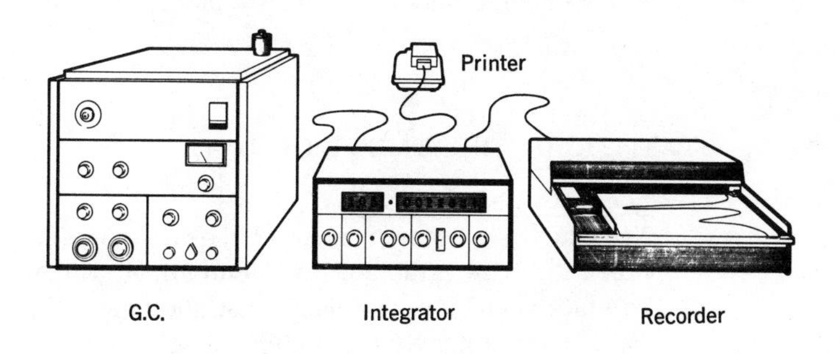

FIGURE VII-12—EXPERIMENTAL SYSTEM

Ten replicate samples were run to obtain the data in Table VII-7. The table summarizes the precision for the entire sample and the time required to work up a trace.

TABLE VII-4—PRECISION OF INTEGRATION METHODS

	Planimeter	Triangulation	H x W 1/2 H	Cut & Weigh	Disc	Digital
Time/Trace - Minutes	45-60	45-60	50-60	100-200	15-30	5-10
Precision σ rel	4.06%	4.06%	2.58%	1.74%	1.29%	0.44%

According to our study, the techniques in increasing precision are planimetry, triangulation, height x width, cutting and weighing, disc integrator, and electronic digital integrator. Cutting and weighing is the slowest method with planimetry, triangulation, and height x width also being time consuming. Disc is fast and the electronic digital integrator very fast. In terms of precision, it appears that height x width, or cutting and weighing are better than planimetry or triangulation. Disc integration is more precise than the manual methods and the electronic digital is considerably more precise than any other method.

D. STATISTICAL TREATMENT OF DATA

1. Accuracy and precision

Accuracy is a measurement of the difference between the true value and the determined values. To state accuracy, one must know the true value, i.e., from a gravimetric blending of standards.

If the true value is known, the difference between the true and measured value is the error. Usually error is expressed as relative error, i.e., $100 \times \frac{\text{error}}{\text{true value}}$.

The accuracy of a measurement is given by stating the relative error. In those cases in which the true value is not known, it is necessary to express the exactness of the measurement in another way.

Precision is one way to express exactness. One obtains the average of a number of measurements and finds the difference (deviation) of each value from the average value. The magnitude of the deviations is a measure of the precision of measurement.

Relative deviation is $100 \times \frac{\text{deviation}}{\text{average}}$

Accuracy is difficult without good precision. Precision however does not insure accuracy. Precision with calibration gives accuracy.

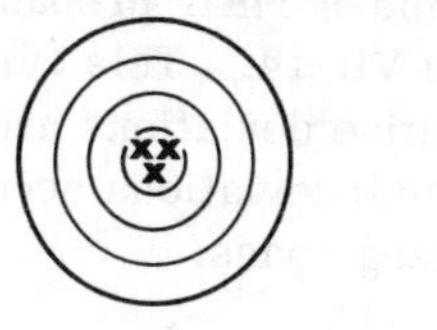

ACCURACY PRECISION

FIGURE VII-13—ACCURACY AND PRECISION

2. Errors

Errors can be divided into two classes: determinate errors, and indeterminate errors. Determinate errors are those errors whose cause and magnitude can be determined. Since they can be determined, correction factors can be applied. Some determinate errors found in G. C. are:

a. Difference in response between different detectors.

b. Change in response of a detector due to temperature, flow, or current changes.

c. Contamination of the sample before injection.

d. Loss of sample or fractionation during injection or on the column.

e. Use of wrong calibration curve.

f. Calculation errors.

g. Taking the wrong sample.

h. Prejudice - changing a result because the operator feels he knows the proper answer.

This list points out the need to watch for determinate errors, and the need to determine their magnitude and apply the proper correction.

Indeterminate errors are random errors which cannot be eliminated. The distribution of these errors follows the normal probability curve as shown in Figure VII-14. This curve shows that positive and negative deviations are equally probable and that small deviations occur much more frequently than large ones.

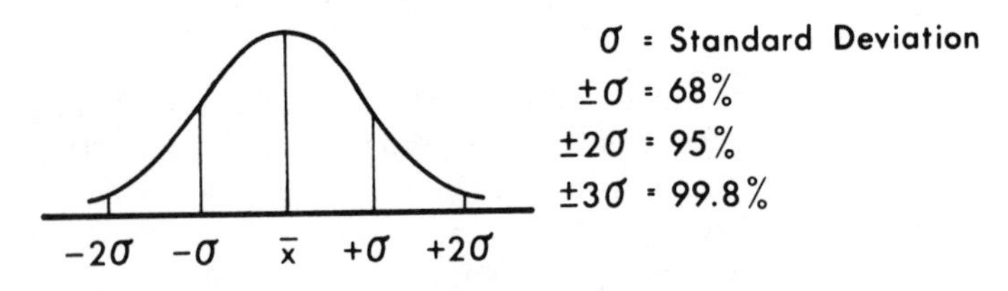

FIGURE VII-14—NORMAL PROBABILITY CURVE

The area on each side of the average can be used to determine the probability that indeterminate errors will occur under similar conditions. The probability of an error falling within the area is as follows:

$\overline{X} \pm 1\sigma$ is 68%; $\overline{X} \pm 2\sigma$ is 95%; $\overline{X} \pm 3\sigma$ 99.8%.

In reporting data the preferred method is to state average ± 1 standard deviation. This means there is a 2/3 probability that future similar measurements will lie within this range.

3. Average and standard deviation

It is recommended that the average value and one standard deviation be used to express G. C. precision.

a. Arithmetic average: all measurements are added together and divided by the total number of measurements. This is a rapid and simple calculation, but it cannot express extreme values. The spread of data can best be expressed by standard deviation.

Example: if two chemists analyzed the benzene content in a sample, it is possible that one chemist has a better and more reproducible technique. Even though the averages are the same, it is obvious that chemist B has a better technique.

TABLE VII-5—FAILURE OF AVERAGE TO SHOW SPREAD

Run No.	Chemist A	Chemist B
1	10.0%	10.2%
2	12.0%	10.6%
3	9.0%	9.8%
4	11.0%	10.1%
5	8.0%	9.3%
Average	$\frac{50.0}{5} = 10.0\%$	$\frac{50.0}{5} = 10.0\%$

b. Standard deviation: this is difficult to calculate, but expresses precision better than other calculations. It expresses the scatter of the measurement around the average value and can be used to set confidence levels on future similar measurements. For Table VII-5, chemist A has a σ of 1.58; chemist B has a σ of 0.48.

$$\sigma = \text{standard deviation} = \sqrt{\frac{\Sigma (x - \bar{x})^2}{N - 1}}$$

$$\sigma_{rel} = \frac{\sigma}{\bar{x}} \times 100$$

where:

N = number of measurements

X = measured values

$\bar{X}$ = Arithmetic mean

Σ = symbol meaning "sum of"

The arithmetic mean ($\bar{X}$) is calculated. The absolute difference between $\bar{X}$ and each measurement is squared and these values summed. The total is divided by N-1, and the square root of the result is σ, the standard deviation.

Example: a beginner in gas chromatography wished to measure the precision of syringe injection technique. He injected 0.5μliter ten times and measured the peak heights, X, in mm. The problem is to calculate the average, standard deviation, and draw conclusions.

We conclude that if the student were to inject another sample, there is a 2/3 probability that the peak height will be between 149.4 and 143.8 (146.6 $\pm$ 2.8). His relative standard deviation is 1.9% which is reasonable

TABLE VII-6—CALCULATION OF STANDARD DEVIATION

N	X	$(X - \overline{X})$	$(X - \overline{X})^2$
1	142.1 mm	142.1 - 146.6 = 4.5	20.25
2	147.0 mm	147.0 - 146.6 = 0.4	0.16
3	146.2 mm	146.2 - 146.6 = 0.4	0.16
4	145.2 mm	145.2 - 146.6 = 1.4	1.96
5	143.8 mm	143.8 - 146.6 = 2.8	7.84
6	146.2 mm	146.2 - 146.6 = 0.4	0.16
7	147.3 mm	147.3 - 146.6 = 0.7	0.49
8	150.3 mm	150.3 - 146.6 = 3.7	13.69
9	145.9 mm	145.9 - 146.6 = 0.7	0.49
10	151.8 mm	151.8 - 146.6 = 5.2	27.04
TOTAL	1465.8		72.24

$$\overline{X} = \frac{1465.8}{10} = 146.6 \qquad \sigma_{abs} = \sqrt{\frac{72.24}{10-1}} = 2.83$$

Peak height = 146.6 mm $\pm$ 2.83 mm

$$\sigma_{rel} = \frac{\sigma_{abs}}{\overline{X}} \times 100 = \frac{2.83}{146.6} \times 100 = 1.9\%$$

for a beginner. Experienced workers easily have 1% and less relative standard deviation.

c. Instant standard deviation[6]: a quick and reliable method to estimate standard deviation consists of simple multiplication of the range by a factor determined by the number of measurements made. Table VII-7 gives the proper factor to use and also shows the reliability which tells how good this estimate is as compared to the calculated standard deviation. The range is the difference between the largest and smallest measurement.

TABLE VII-7—FACTORS FOR INSTANT STANDARD DEVIATION

No. of Determinations	Range Multiplier	Reliability
2	0.886	100%
3	0.591	99%
4	0.486	98%
5	0.430	96%
6	0.395	93%
7	0.370	91%
8	0.350	89%
9	0.337	87%
10%	0.325	85%

Table VII-8 compares the instant standard deviation, σ_S, against the classical standard deviation, σ, on two different sets of data. Agreement between σ and σ_S is fairly good.

TABLE VII-8—σ_S VS. σ

	No. 1	No. 2
	17.65	4.51
	17.83	4.49
	17.92	4.59
	17.63	4.53
	17.71	4.46
Average	17.75	4.52
Range	0.29	0.13
Instant σ_S	0.125	0.056
Calculated σ	0.124	0.049

Example: calculate the instant standard deviation in data #2 of Table VII-8.

Range = highest value - lowest value

= 4.59 - 4.46 = 0.13

Number of determinations = s therefore multiplier

Range = highest value - lowest value

= 4.59 - 4.46 = 0.13

Number of determinations = 5, therefore multiplier = 0.43 (see Table VII-7):

Instant standard deviation, σ_s = Range x multiplier

= 0.13 x 0.43

= 0.056

BIBLIOGRAPHY

1. Desty, D.H., Geach, C.J., and Goldup, A., Gas Chromatography 1960, ed. by R.P.W. Scott, Butterworths, 1960, p. 46.
2. Condon, R.D., Scholly, P.R., and Averill, W.A., Gas Chromatography 1960, ed. by. R.P.W. Scott, Butterworths, 1960, p. 30.
3. Halasz, I., Annual General Meeting of the Gas Chromatography Discussion Group, Birmingham, April 1961.
4. Dietz, W.A., Journal of Gas Chromatography, Vol. 5, 68 (1967).
5. Rosie, D.M., et. al., Analytical Chemistry 31, 230 (1959).
6. Aerograph Research Notes, Fall Issue, 1965.

VIII. CHROMATOGRAM INTERPRETATION

The following pages have been inserted in order to help the chromatographer to better interpret the multitude of different peak shapes encountered in gas chromatography. The various chromatograms obtained are the result of our experiences combined with a thorough literature search. We encourage contributions and suggestions.

The injection point on each chromatogram is shown by a tick mark on the baseline as shown below. The time axis runs from left to right (see arrow).

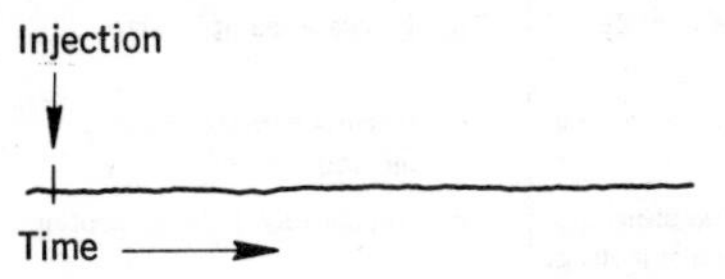

SYMPTOM	POSSIBLE CAUSE	CHECKS AND/OR REMEDY
1. No peaks	1a. Detector (or electrometer) power switch OFF.	1a. Turn detector (or electrometer) switch ON and adjust to desired sensitivity level.
	b. No carrier gas flow.	b. Turn carrier gas flow ON and adjust to proper setting. If carrier lines are obstructed, remove obstruction. Replace carrier-gas tank if empty.
	c. Recorder improperly connected.	c. Connect recorder as described in recorder and/or instrument manual. Remove any jumper lines connecting either recorder input connection to ground or shield.
	d. Recorder defective.	d. See recorder instruction manual for troubleshooting procedures.
	e. Injector temperature too cold. Sample not being vaporized.	e. Increase injector temperature. Check with volatile sample such as air or acetone.
	f. Hypodermic syringe leaking or plugged up.	f. Replace syringe.
	g. Injector septum leaking.	g. Replace injector septum.
	h. Column connections loose.	h. Tighten column connections.

SYMPTOM	POSSIBLE CAUSE	CHECKS AND/OR REMEDY
	i. Flame out (FID only).	i. Inspect flame; feel if hot, light if necessary.
	j. No cell voltage being applied to detector (all ionization detectors).	j. Place CELL VOLTAGE in ON position. Also check for bad detector cables. Measure voltage with voltmeter per instruction manual.
	k. Column temperature too cold. Sample condensing on column.	k. Increase column temperature.
2. Poor sensitivity with normal retention time.	2a. Attenuation too high.	2a. Reduce attenuation. Place RANGE switch in lower setting (ionization detectors only).
	b. Insufficient sample size.	b. Increase sample size.
	c. Poor sample injection technique.	c. Review sample injection techniques.
	d. Syringe or septum leaking when injecting.	d. Replace syringe or septum.
	e. Carrier gas leak.	e. Find and correct leak.
	f. Thermal conductivity response low.	f. Use higher filament current; low detector temperature; different carrier gas, higher resistance filament.
	g. FID response low.	g. Try more hydrogen, more air, faster flow rate; position collector electrode closer to flame, clean detector.
3. Poor sensitivity with increased retention time.	3a. Carrier-gas flow rate too low.	3a. Increase carrier gas flow. If carrier gas lines are obstructed, locate and remove obstruction.
	b. Flow leaks downstream of injector.	b. Locate flow leak and correct.
	c. Injector septum leaking continuously.	c. Replace injector septum.

SYMPTOM	POSSIBLE CAUSE	CHECKS AND/OR REMEDY
4. Negative peaks. (I) 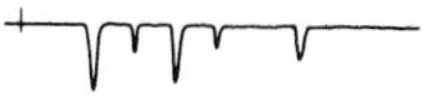 (II) 	4a. Recorder improperly connected. Input leads reversed.	4a. Connect recorder as described in recorder and/or instrument manual.
	b. Sample injected in wrong column.	b. Inject sample in proper column.
	c. MODE switch in wrong position. (Ionization detectors).	c. Insure MODE switch is in correct position for column being used as analytical column.
	d. POLARITY switch in wrong position (thermal conductivity detector).	d. Change POLARITY switch.
5. Irregular baseline drift when operating isothermally. 	5a. Poor instrument location.	5a. Move instrument to a different location. Instrument should not be placed directly under heater or air conditioner blower, or any other place where it is subject to excessive drafts and ambient temperature changes.
	b. Instrument not properly grounded.	b. Insure instrument and recorder connected to good earth ground.
	c. Column packing bleeding.	c. Stabilize column as outlined in instrument manual. Some columns are impossible to stabilize well at the desired operating conditions. These columns will always produce some baseline drift, particularly when operating at high sensitivity conditions.
	d. Carrier gas leak.	d. Locate leak and correct.
	e. Crossover lines to detector oven contaminated.	e. Clean crossover lines.
	f. Detector block contaminated (TC detectors.)	f. Clean detector block.
	g. Detector base contaminated (Ionization detectors).	g. Clean detector base. See instrument manual.
	h. Detector-oven temperature not stable (TC detector).	h. Make sure plugs are installed in marinite oven lid if ionization detectors are not installed; also, make sure there are no gaps in the detector oven insulation which will allow room air to enter detector oven.

SYMPTOM	POSSIBLE CAUSE	CHECKS AND/OR REMEDY
	i. Poor carrier gas regulation.	i. Check carrier-gas regulator and flow controllers to ensure proper operation. Make sure carrier gas tank has sufficient pressure.
	j. Poor H_2 or air regulation (FID only).	j. Check H_2 and air flow to ensure proper flow rate and regulation.
	k. Detector filaments defective (TC detector only).	k. Replace TC detector assembly or filaments.
	l. Electrometer defective (Ionization detectors).	l. See instrument manual on electrometer troubleshooting.
	m. Recorder defective.	m. Disconnect input leads to recorder and short the recorder input with a piece of wire. If drift continues, recorder is defective. See recorder manual. Especially, check recorder gain adjustment, damping adjustment, ground connection and amplifier tubes.
	n. TC Power Supply defective.	n. Replace Power Supply.
6. Sinusoidal baseline drift.	6a. Detector oven insulation faulty.	6a. Make sure insulation is correctly installed or replaced.
	b. Detector oven temperature controller defective.	b. Replace detector oven temperature controller, and/or temperature sensing probe.
	c. Column oven temperature defective.	c. Replace column oven temperature control module and/or temperature sensing probe.
	d. OVEN TEMP. °C control on main control panel set too low.	d. Set OVEN TEMP. °C control to higher setting. Must be set higher than highest desired operating temperature of the column oven.
	e. Carrier gas flow regulator defective.	e. Replace carrier-gas flow regulator; sometimes higher pressure provides better control.
	f. Carrier gas tank pressure too low to allow regulator to control properly.	f. Replace carrier-gas tank.

SYMPTOM	POSSIBLE CAUSE	CHECKS AND/OR REMEDY
7. Constant baseline drift in one direction when operating isothermally. 	7a. Detector temperature increasing (decreasing).	7a. Allow sufficient time for detector to stabilize after changing its temperature. Particularly important with TC detector. Detector block will lag the indicated temperature somewhat because of its large mass.
	b. Flow leak down stream of column effluent end (TC detector only).	b. A very small diffusion leak will allow a small amount of air to enter the detector at a constant rate. This in turn will oxidize the effected elements at a constant rate while slowly changing their resistance. Locate leak and correct. These are often very slight leaks, and difficult to find. Use high carrier gas pressure (60-70 psig) if necessary.
	c. Defective detector filaments (TC detector).	c. Replace detector or filaments.
	d. Defective power supply (TC detector).	d. Replace power supply.
	e. Defective electrometer (Ionization detectors).	e. See instrument manual on electrometer troubleshooting.
8. Rising baseline when temperature programming. 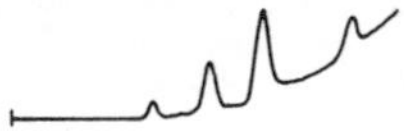	8a. Increase in column "bleed" when temperature rises.	8a. Use reference column (dual column operation) and condition column. Use less liquid phase and lower temperature.
	b. Column flow rates not balanced for optimum performance.	b. Balance column flow rates as outlined in instruction manual.
	c. Column(s) contaminated.	c. Recondition columns.
9. Irregular baseline shifting when temperature programming. 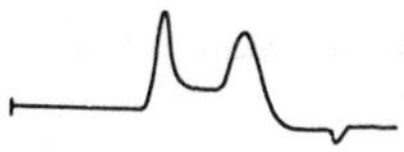	9a. Excessive column "bleeding" from well conditioned columns.	9a. Use less liquid phase and low temperatures. Use different columns. Some packing materials are highly volatile so that they cannot be temperature programmed without difficulty. This will occur even on well conditioned columns where the carrier gas flow rates have been carefully optimised.

SYMPTOM	POSSIBLE CAUSE	CHECKS AND/OR REMEDY
	b. Columns not properly conditioned.	b. Condition columns as outlined in instruction manual.
	c. Carrier gas flow rates not balanced for optimum performance.	c. Balance column flow rates as outlined in instruction manual.
	d. Column(s) contaminated.	d. Recondition columns.
10. Baseline "stepping". Baseline does not return to zero, attenuation is incorrect, and peaks are "flat topped". Recorder pen easily moved upscale or downscale with finger pressure.	10a. Recorder gain and/or damping control improperly adjusted.	10a. Adjust recorder gain and/or damping control. Refer to recorder instruction manual. When adjusted properly recorder pen cannot be moved easily with finger pressure.
(I) 	b. Instrument and/or recorder not properly grounded.	b. Insure instrument and recorder are connected to good earth ground.
(II) 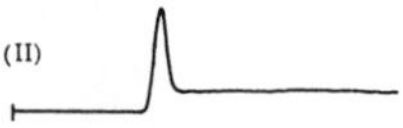	c. Low level a-c signal being fed to recorder.	c. Install filter capacitor (about 0.25uf 150 vdc) from either + or - input of recorder to ground. Determine by trial and error which position (or both) works best. Do not connect capacitor across recorder input (i.e. from + to - connection).
	d. Physical damage to filaments (TC). Sample may contain high halogen, oxygen or sulfur (chromatogram II).	d. Replace TC cell.
11. Baseline cannot be zeroed.	11a. Adjustable zero on recorder improperly set.	11a. Short recorder input with piece of wire and adjust zero. See recorder instruction manual.
	b. Detector filaments out of balance. (TC detector).	b. Replace detector.
	c. Defective power supply (TC detector).	c. Replace power supply.
	d. Excessive signal from column "bleed" (especially FID).	d. Use different column with less "bleed".
	e. Dirty detector (FID and EC).	e. Clean detector base and head assemblies.

SYMPTOM	POSSIBLE CAUSE	CHECKS AND/OR REMEDY
	f. Defective electrometer.	f. See instrument manual on electrometer troubleshooting.
	g. Recorder improperly connected.	g. Connect recorder as described in recorder and/or instrument manual. Remove any jumper lines connecting either recorder input connection to ground or shield.
	h. Recorder defective.	h. See recorder instructions.
12. Sharp "spiking" at irregular intervals.	12a. Quick atmospheric pressure changes from opening and closing doors, blowers, etc.	12a. Locate instrument to minimize problem. Also do not locate under heater or air conditioner blowers.
	b. Dust particles or other foreign material burned in flame (FID only).	b. Take care to keep detector chamber free of glass wool, maranite, molecular sieve (from air filter) etc. dust particles. Blow out or vacuum detector to remove dust.
	c. Dirty insulators and/or connectors (Ionization detectors).	c. Clean insulators and connectors with residue free solvent. Do not touch with bare fingers after cleaning.
	d. Power supply defective (TC detector).	d. Replace power supply.
	e. Defective electrometer (Ionization detectors).	e. See electrometer troubleshooting in instrument manual.
	f. High line voltage fluctuations.	f. Use separate electrical outlet; use stabilized transformer.
13. Short spikes or peaks at regular intervals.	13a. Condensation in flow lines causing carrier gas to bubble through.	13a. Heat lines to remove condensation or blow out lines.
	b. Bubble meter with high liquid level attached to detector exit line (TC detector).	b. Remove soap-bubble flowmeter tube from exit line.
	c. Water condensation in hydrogen line coming from Model 650 Hydrogen Generator (FID only).	c. Remove water from line and replace Hydrogen filter.
	d. High line voltage fluctuations.	d. Use separate electrical outlet; use stabilized transformer.

SYMPTOM	POSSIBLE CAUSE	CHECKS AND/OR REMEDY
14. High background signal (noise).	**14a. Contaminated column or excessive bleed from column.**	**14a. Recondition column.**
	b. Contaminated carrier gas.	**b. Replace or regenerate carrier gas filter. Regenerate filter by heating to about 175-200°C and purging overnight with dry nitrogen.**
	c. Carrier gas flow rate too high.	**c. Reduce carrier gas flow rate.**
	d. Carrier gas flow leak.	**d. Locate leak and correct.**
	e. Loose connections.	**e. Make sure all interconnecting plug and screw connections are tight. Make sure modules are properly seated in their plug-in connectors.**
	f. Bad ground connection.	**f. Insure all ground connections are tight and connected to a good earth ground.**
	g. Dirty switches.	**g. Locate dirty switch, spray with a contact cleaner and rotate switch through its positions several times.**
	h. Dirty recorder slidewire.	**h. Clean recorder slidewire. See recorder manual. If slidewire is dirty noise will always appear at same area or areas on chart and noise level will generally remain constant regardless of attenuation.**
	i. Defective recorder.	**i. Short recorder input with piece of wire. If noise continues, check recorder. See recorder manual.**
	j. Dirty injector.	**j. Clean injector tube and replace septum.**
	k. Dirty crossover block from column oven to detector oven.	**k. Clean crossover block.**
	l. Dirty detector (TC detector).	**l. Clean detector block.**
	m. Defective detector filaments (TC detector).	**m. Replace detector assembly.**
	n. Defective bridge circuit and/or power supply (TC detector).	**n. Replace TC module.**
	o. Hydrogen flow rate too high or too low (FID detector).	**o. Adjust hydrogen flow rate to proper level.**
	p. Air flow too high or too low (FID detector).	**p. Adjust air flow rate to proper level.**

SYMPTOM	POSSIBLE CAUSE	CHECKS AND/OR REMEDY
	q. Air or hydrogen contaminated (FID detector).	q. Replace or regenerate air and hydrogen filters.
	r. Water condensing inside flame detector shell (FID detector).	r. Raise detector temperature to approximately 100°C to eliminate condensation.
	s. Defective detector cables (Ionization detectors).	s. Replace detector cables.
	t. Dirty detector insulators (Ionization detector).	t. Clean insulators thoroughly with residue free solvent. Do not touch clean insulators with fingers.
	u. Dirty detector base and/or shell (Ionization detectors).	u. Clean detector base and shell. Use ultrasonic bath if possible.
	v. Leak around flame tip (FID)	v. Tighten nut or replace washer.
15. Tailing peaks. 	15a. Injector temperature too high or too low.	15a. Readjust injector temperature.
	b. Injector tube dirty (sample or septum residue).	b. Clean injector tube with solvent and pipe cleaner. If necessary replace entire injector.
	c. Column-oven temperature too low.	c. Increase column-oven temperature. Do not exceed recommended maximum temperature for packing material.
	d. Poor sample injection technique.	d. Review Syringe Handling Techniques.
	e. Wrong column. Interaction between sample material and column solid support and/or liquid phase.	e. Use different column. Use higher % liquid phase, more polar liquid phase or more inert solid support.
16. Leading peaks 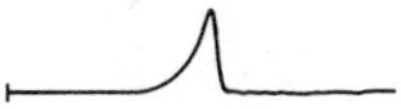	16a. Column overloaded. Sample size too large for column diameter and length.	16a. Decrease sample size.
	b. Sample condensed in system.	b. Insure injector and detector temperatures are correct.
	c. Poor sample injection technique.	c. Review Syringe Handling Techniques.

SYMPTOM	POSSIBLE CAUSE	CHECKS AND/OR REMEDY
17. Unresolved peaks. (I) (II)	17a. Column oven temperature too high.	17a. Lower column oven temperature.
	b. Column too short.	b. Use longer column.
	c. Liquid phase has "baked" off column support material.	c. Replace column.
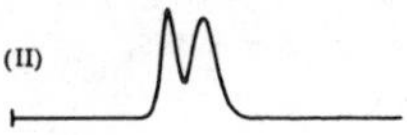	d. Wrong column. Incorrect choice of liquid phase and/or solid support.	d. Use different column.
	e. Carrier gas flow rate too high.	e. Reduce carrier gas flow rate.
	f. Poor injection technique.	f. Review Syringe Handling Techniques.
18. Round top peaks. 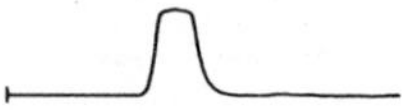	18a. Operating beyond linear dynamic range of detector output (especially noticeable with EC detector).	18a. Reduce sample size.
	b. Recorder gain too low.	b. Adjust recorder gain to proper setting. See recorder manual.
19. Square top peaks. 	19a. Electrometer input tube saturated (Ionization detectors).	19a. Reduce sample size. Switch electrometer RANGE control to higher setting.
	b. Recorder slidewire defective or mechanical binding of drive system.	b. Check recorder operation on a separate instrument or millivolt source. See recorder manual.

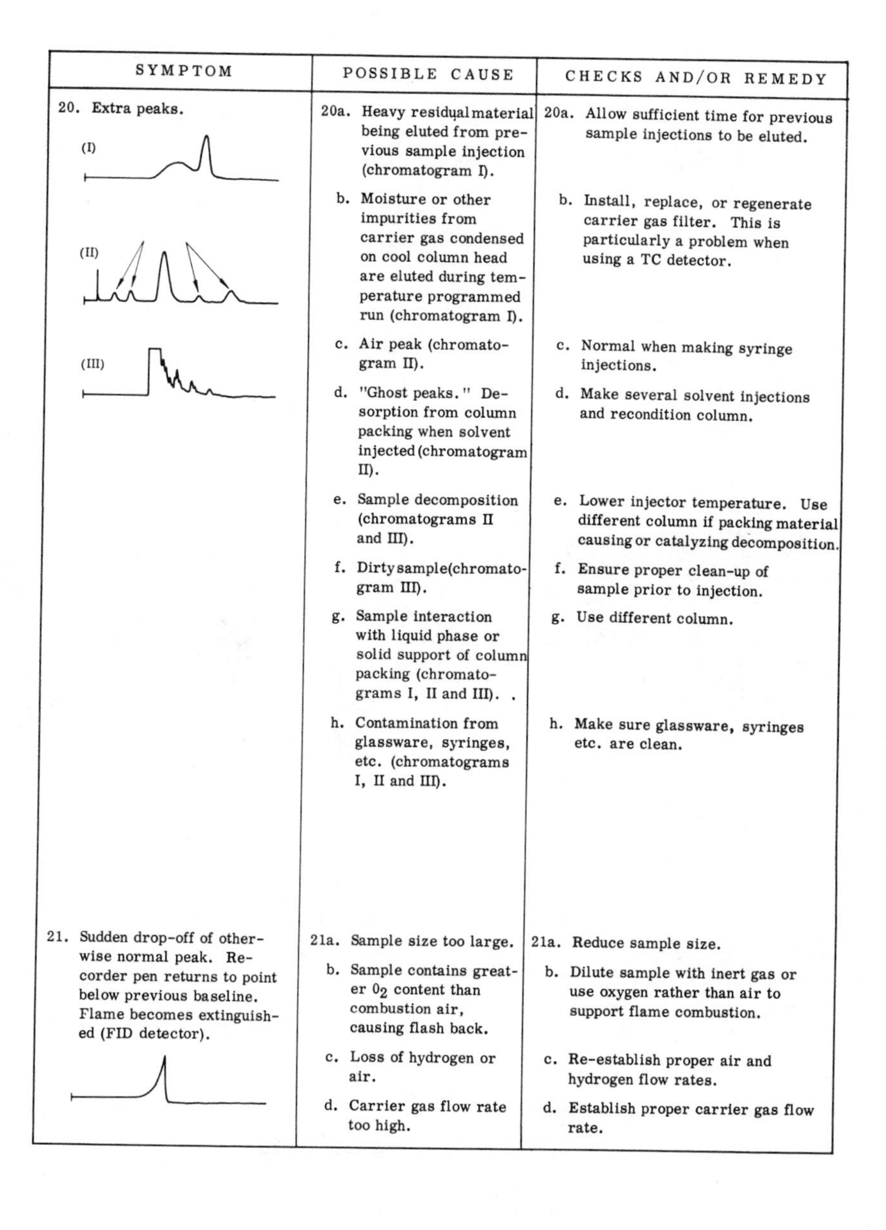

SYMPTOM	POSSIBLE CAUSE	CHECKS AND/OR REMEDY
20. Extra peaks. (I) (II) (III)	20a. Heavy residual material being eluted from previous sample injection (chromatogram I).	20a. Allow sufficient time for previous sample injections to be eluted.
	b. Moisture or other impurities from carrier gas condensed on cool column head are eluted during temperature programmed run (chromatogram I).	b. Install, replace, or regenerate carrier gas filter. This is particularly a problem when using a TC detector.
	c. Air peak (chromatogram II).	c. Normal when making syringe injections.
	d. "Ghost peaks." Desorption from column packing when solvent injected (chromatogram II).	d. Make several solvent injections and recondition column.
	e. Sample decomposition (chromatograms II and III).	e. Lower injector temperature. Use different column if packing material causing or catalyzing decomposition.
	f. Dirty sample (chromatogram III).	f. Ensure proper clean-up of sample prior to injection.
	g. Sample interaction with liquid phase or solid support of column packing (chromatograms I, II and III).	g. Use different column.
	h. Contamination from glassware, syringes, etc. (chromatograms I, II and III).	h. Make sure glassware, syringes etc. are clean.
21. Sudden drop-off of otherwise normal peak. Recorder pen returns to point below previous baseline. Flame becomes extinguished (FID detector).	21a. Sample size too large.	21a. Reduce sample size.
	b. Sample contains greater O_2 content than combustion air, causing flash back.	b. Dilute sample with inert gas or use oxygen rather than air to support flame combustion.
	c. Loss of hydrogen or air.	c. Re-establish proper air and hydrogen flow rates.
	d. Carrier gas flow rate too high.	d. Establish proper carrier gas flow rate.

SYMPTOM	POSSIBLE CAUSE	CHECKS AND/OR REMEDY
	e. Flame tip fouled.	e. Clean or replace flame tip.
	f. Hydrogen generator "kick-off" due to back pressure surge.	f. Reset hydrogen generator. If it "kicks-off" again look for obstruction in hydrogen flow system. Clean H_2 flow system and reset generator
22. Negative dips after peaks (EC detector). 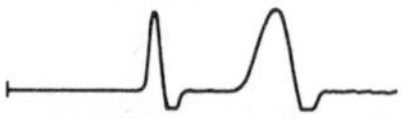	22a. Contaminated detector.	22a. Clean detector.
23. "Screw cap effect" with EC detector. Large broad tailing peak. 	23a. Sample container cap liner partially dissolved in sample solvent.	23a. Line sample container cap with metal foil, or use glass or polyethylene stoppered containers.

IX. TEMPERATURE PROGRAMMING

A. INTRODUCTION

Temperature programming is the controlled change of column temperature during an analysis. It is used to improve, simplify, or accelerate the separation, identification, and determination of sample components. We shall restrict ourselves to discussing linear programs of temperature increases over the entire column length. Nonlinear programs, temperature decreases and temperature gradients on the column have limited advantages.

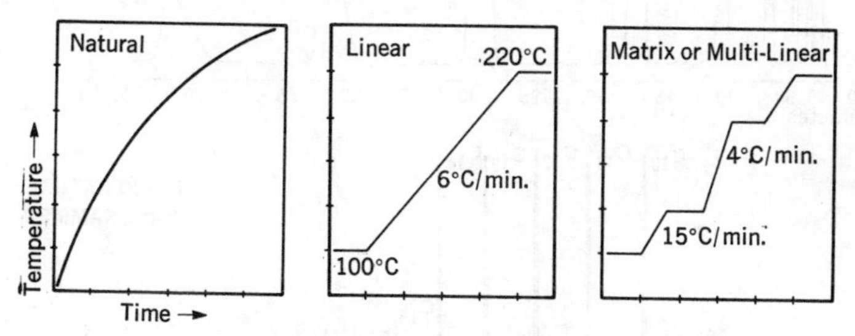

FIGURE IX-1—NATURAL, LINEAR, AND MATRIX PROGRAMS

Programmed temperature gas chromatography (PTGC) is a logical extension of the isothermal method and resulted from the limitations of the constant temperature technique for the analysis of complex mixtures and wide boiling range samples.

As shown in Figure IX-2, isothermal operation limits gas chromatographic analysis to a narrow boiling sample. At constant temperature the early peaks, representing low boiling components, emerge so rapidly that sharp overlapping peaks result while higher boiling materials emerge as flat, immeasurable peaks. In some cases, high boiling components are not eluted and may appear in a later analysis as baseline noise or "ghost" peaks which cannot be explained. Conditions were: Sample of Normal Paraffins, 20 feet by 1/16 inch column, 3% Apiezon L on 100/120 mesh Aeropak 30 at 150°C, 10 ml/minute He.

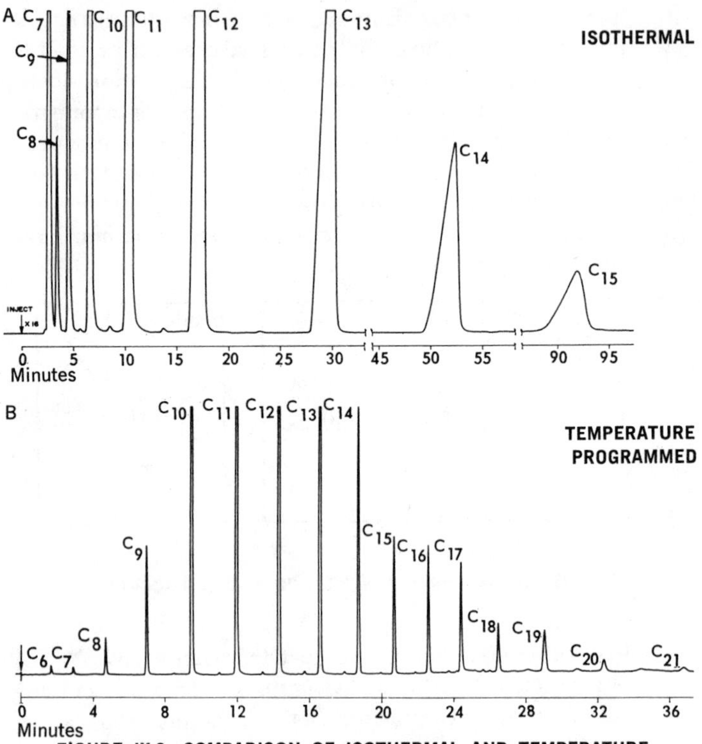

FIGURE IX-2—COMPARISON OF ISOTHERMAL AND TEMPERATURE PROGRAMMED CHROMATOGRAMS

With temperature programming, a lower initial temperature is used and the early peaks are well resolved. As the temperature increases, each higher boiling component is "pushed" out by the rising temperature, High boiling compounds are eluted earlier and as sharp peaks, similar in shape to the early peaks. Trace components emerge as sharp peaks and can be more easily distinguished from the baseline. Total analysis time is much shorter.

Temperature programming allows the proper selection of a temperature which will result in well-resolved, nicely shaped peaks, and a total analysis time shorter than isothermal operation. Temperature programming is simply a means for obtaining, automatically, the ideal temperature range for the separation of each narrow boiling point fraction or component. Each sample automatically selects an ideal temperature in which to migrate and separate within the column. Prior to reaching this ideal temperature range, each substance is "frozen" or condensed at the head of the column,

TABLE IX-1—COMPARISON OF ISOTHERMAL AND PROGRAMMED TEMPERATURE GAS CHROMATOGRAPHY

ASPECT	ISOTHERMAL	PTGC
Boiling range of sample	Limited to 100°C	80°C to 400°C
Precision of Peak Measurement	Varies with peak shape	Little variation
Detection Limits	Varies with peak shape	Little variation
Sample Injection	Must be fast	Need not be fast
Stationary Phase	Wide choice	Restricted choice
Purity of Carrier Gas	Not critical	High purity essential
Separate ovens for Column & Detector	Convenient	Essential
Flow Control	Constant pressure sufficient	Differential flow controller required.

waiting for its turn to be separated at a higher temperature. There are no strange or unusual effects originating in temperature change itself. As Giddings says, "*The heating process may be considered merely as a mechanism for obtaining a range of temperature in proper sequence with no direct effect upon the separation obtained* [1]."

The decision to use PTGC is based on consideration of the boiling points of the sample components. Generally, if the range of boiling points is 100°C or more, programming is advisable. Programming also helps in preparative scale, trace analysis, and gas solid chromatography as will be discussed later.

B. PTGC REQUIREMENTS

Several requirements must be met for temperature programmed operation:

1. Separate heaters for injection port, column oven, and detector.
2. A programmer with a range of programming rates (1/4°C to 20°C per minute).
3. A low mass oven.
4. A suitable liquid phase.
5. A differential flow controller.
6. Pure, dry carrier gas.

These features are essential for temperature programming. Though not essential, they will also provide better, more stable performance for isothermal operation.

These requirements will be discussed in more detail to illustrate their importance to the temperature programming technique.

1. <u>Separate heaters</u>
   The injection port, column oven, and detector oven should be controlled by separate heaters.

Each should be well insulated from the other. The purpose of temperature programming is to allow rapid heating and cooling of the column oven. It is not desirable that the injector port or detector oven change temperature during the analysis. Particularly with thermal conductivity detectors, it is essential that the detector block temperature remain as constant as possible to avoid baseline drift and detector response changes.

The flame detector is quite good for high temperature programmed operation since it is not sensitive to small temperature changes. it is necessary to keep the electrical leads from the detector to the electrometer well insulated from the hot detector.

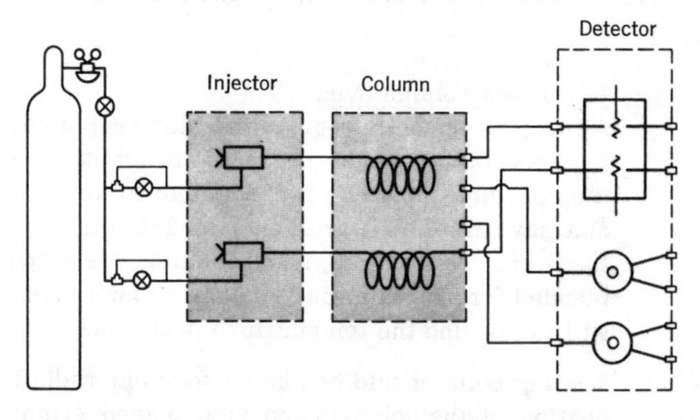

FIGURE IX-3—SEPARATE HEATED ZONES REQUIRED FOR TPGC

2. <u>Programmer</u>

   It is necessary to have some mechanism, either mechanical or electronic, which can precisely reproduce a range of programming rates (1/4°C to 20°C per minute). These are commercially available.

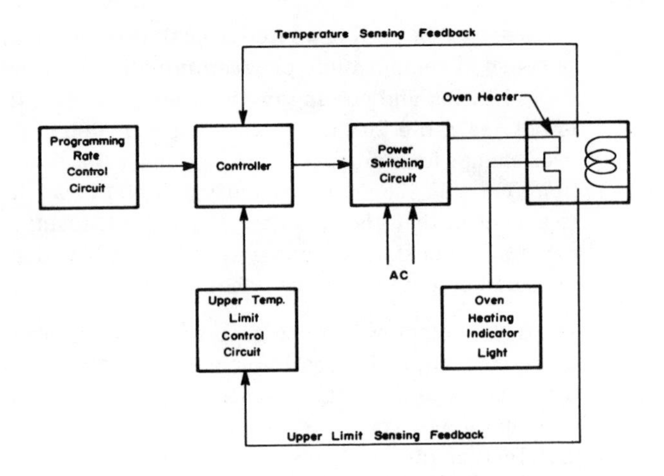

FIGURE IX-4—SCHEMATIC OF A TEMPERATURE PROGRAMMER

3. <u>Low mass column oven</u>
   A low mass oven is required to allow rapid heating and cooling of the column. In addition, the temperature should be reproducible to within 1°C and have minimum gradients of 2°C throughout the oven. Thin walled, short columns are recommended for programming. Large column masses will lag behind the temperature program.

   Heating coils should be placed to avoid radiation heating of the columns and also placed symmetrically to avoid temperature gradients.

   High mass ovens will not cool rapidly. Thin-walled, stainless steel ovens with tight lids and high speed circulating air fans seem to make the best ovens for programming.

4. <u>A suitable liquid phase</u>

   a. Temperature stability
      The liquid phase must be stable at maximum

operating temperature. A commonly used limit is a maximum vapor pressure of 10^{-6}g. liquid phase per milliliter of carrier gas.

Vaporization of the liquid phase is often referred to as *bleeding*. If the column bleeds, it produces noise, a shifting baseline, and changes in column characteristics. The number of high temperature liquid phases is quite limited. Maximum allowable temperatures of some phases used in programming are given in Table IX-2. A complete list of liquid phases and maximum temperatures is to be found in the Appendix.

TABLE IX-2—MAXIMUM COLUMN TEMPERATURE

NON-POLAR PHASES		POLAR PHASES	
SE-30	375°C	Versamid 900	250°C
QF-1	250°C	FFAP	275°C
Apiezon-L	300°C	Carbowax 20M	250°C
SF-96	300°C	EGSS-X	225°C
		STAP	250°C

The use of low liquid loadings and narrow columns help in that lower temperatures are required to elute high boiling peaks. In addition, low loaded narrow columns bleed less since the bleeding is proportional to the grams of liquid phase present. Figure IX-5 shows normal paraffins up to C_{21} eluted from a single 20-foot by 1/16-inch OD packed column containing 3% Apiezon L. Note that the baseline drift is negligible even up to 250°C with a highly sensitive FID.

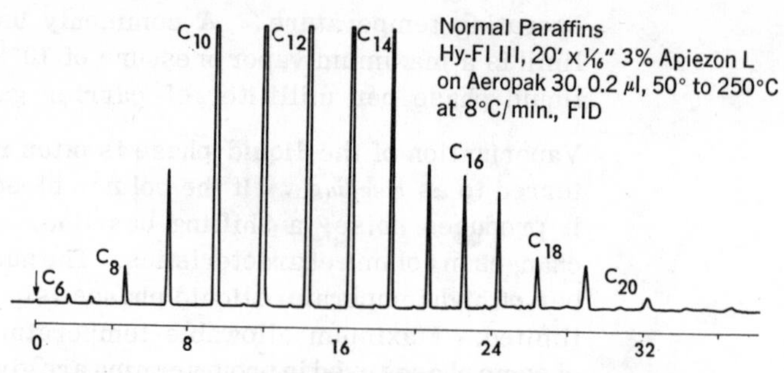

FIGURE IX-5—TEMPERATURE PROGRAMMING USING THIN, LOW-LOADED COLUMNS

Low liquid loadings require the use of small samples and inert support materials. The total amount of liquid phase should be sufficient to provide a column length allowing adequate resolution without requiring temperatures so high as to affect the sample, liquid phase, or detector.

b. Dual columns to compensate bleeding
The use of a dual column gas chromatograph such as illustrated in Figure IX-6 can compensate for the bleed of a single column. The

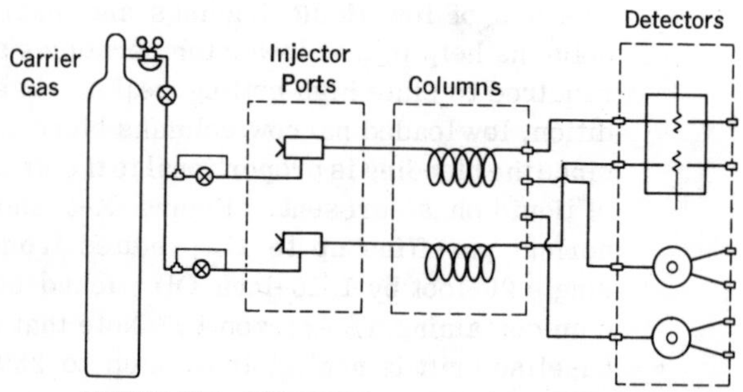

FIGURE IX-6—SCHEMATIC OF A DUAL COLUMN GAS CHROMATOGRAPH

signal from one column is opposed to the signal from the other column. The bleeding is not eliminated, but the difference between the bleed from two matched columns helps maintain a stable baseline to higher operating temperature (see Figure IX-7).

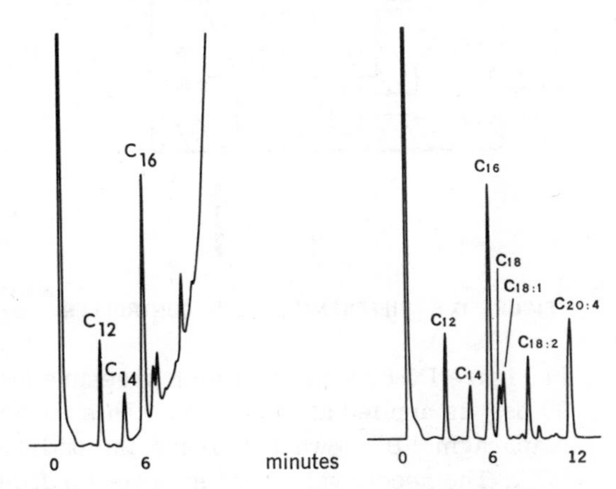

Methyl Esters
Aerograph Model 204-1, 5′ x ⅛″ 20% DEGS
on 60/80 HMDS Chromosorb W
160 to 210°C

FIGURE IX-7—SINGLE AND DUAL COLUMN UNDER PROGRAMMED CONDITIONS

5. Flow controller

A differential flow controller is required to provide a constant carrier gas flow rate during programming. As temperature increases, carrier gases expand, column resistance increases, and flow rate decreases. Under constant pressure conditions (a simple needle valve), the flow rate would decrease and the baseline would drift. Thermal conductivity detectors could not be used with a changing flow rate. The baseline would not be stable and the detector response would change.

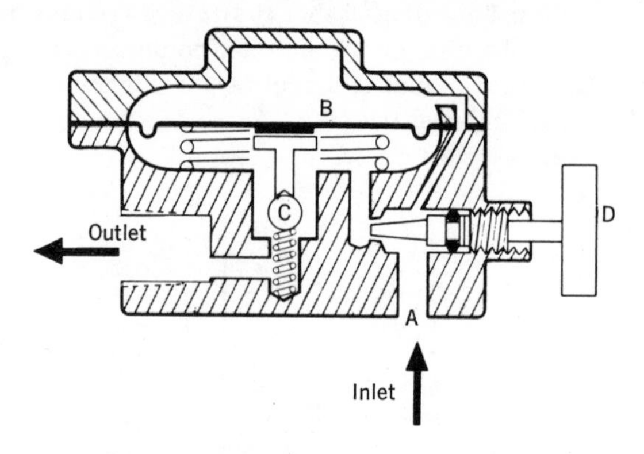

FIGURE IX-8—DIFFERENTIAL FLOW CONTROLLER

In Figure IX-8, a constant high pressure (usually 60 psi) is applied at inlet "A". This pushes the diaphragm "B" down and closes the outlet valve "C". The needle valve "D" is opened sufficiently to increase the upward pressure on the diaphragm and allow some gas to flow through "C" to the column. Changes in the downstream pressure will be compensated for by the spring loaded diaphragm and a constant mass flow results. Stable flow rates can be obtained over a wide range of temperatures[4].

TABLE IX-3—DIFFERENTIAL FLOW CONTROLLER RESULTS

Column Temperature °C	Inlet Pressure cm of Hg	Helium Flow ml/min.
20	26.3	70.7
50	27.9	61.5
100	32.4	60.5
150	35.0	60.6
200	41.2	60.7

6. Pure, dry carrier gas

A Molecular Sieve (5A) filter is recommended for temperature programming. This removes traces of water which produce ghost peaks under programmed conditions. A 2-foot x 1/4-inch column is sufficient to dry one large gas cylinder. Regeneration is easily accomplished by heating at 350°C for several hours with dry nitrogen flowing.

C. VISCOSITY OR LOW TEMPERATURE SUITABILITY

The freezing point or viscosity determines the minimum operating temperature of liquid phases. Many high temperature liquid phases are solids at the low temperatures required for programming. Other liquid phases, while still liquid, have such high viscosities that column efficiency is sacrificed. The loss of efficiency results from excessive band spreading due to the high resistance to mass transfer in the liquid phase (see Chapter III). Admittedly, this efficiency loss will affect only those early eluting peaks with low retention temperatures. If possible, the liquid phase should have viscosities of about one poise or smaller. As Table IX-4 shows, some liquids have very high viscosities at low temperatures[3].

TABLE IX-4—LIQUID PHASE VISCOSITIES AT DIFFERENT TEMPERATURES

	Viscosity (poise)			
Liquids	0.10	0.30	1.0	3.0
Apiezon L	200°C	140°C	95°C	67°C
Glycerol	112°C	81°C	57°C	40°C
Squalane	61°C	24°C	-	-
Tricresyl Phosphate	77°C	38°C	19°C	7°C

D. ELEMENTARY THEORY

Although the application of PTGC is relatively simple, the theory of this technique leads to complex integrals of difficult solution. A good and simple approximation which shows how a solute peak migrates through a chromatographic column has been offered by Giddings[1]. A more thorough treatment of the theory is found in the book by Habgood and Harris[2]. We recommend both references to the eager student, but we shall not discuss the theory in this book. We present some simple operating rules derived from the theory.

1. A column length is chosen on the basis of resolution needed. With packed columns, a typical length is 6 to 10 feet. Longer lengths are not beneficial in programming unless smaller heating rates are used.

2. The initial temperature is chosen much the same as for the isothermal analysis of the lower boiling components. This will normally be lower than the boiling point of the lowest boiling component. One guideline is that it should be 90° cooler than the elution temperature. A lower initial temperature has negligible effect upon the resolution for high boilers since these solutes are essentially "frozen" and thus unaffected by the early parts of the temperature program.

3. The heating rate (β) chosen is a compromise between resolution and analysis speed. The heating rate provides the same function in temperature programmed analysis as the operating temperature does in isothermal analysis. At lower rates, analysis time will be too long for the high boilers and band width deterioration will take place. At high rates, severe loss of resolution will occur. Typical rates for 6 to 10 foot columns of 1/8 and 1/4-inch diameter are $1^{o}C$ to $4^{o}C$ per minute.

4. The flow rate has a relatively small effect on the analysis time as compared to the temperature parameters and is, therefore, relatively unimportant. The flow velocity should be the optimum or higher as determined from the Van Deemter plot.

5. The final temperature chosen should be near the boiling point of the highest boiling component present. Of course, one must adhere to the practical limitations of column substrate volatility. Some of the most widely used subtrates such as SE-30 and Apiezon L are routinely programmed to 300°C.

E. APPLICATIONS OF PTGC

1. Wide boiling range samples

Illustrated in Figure IX-9 is a separation of the C_2 to C_{18} alcohols with one sample, dual columns,

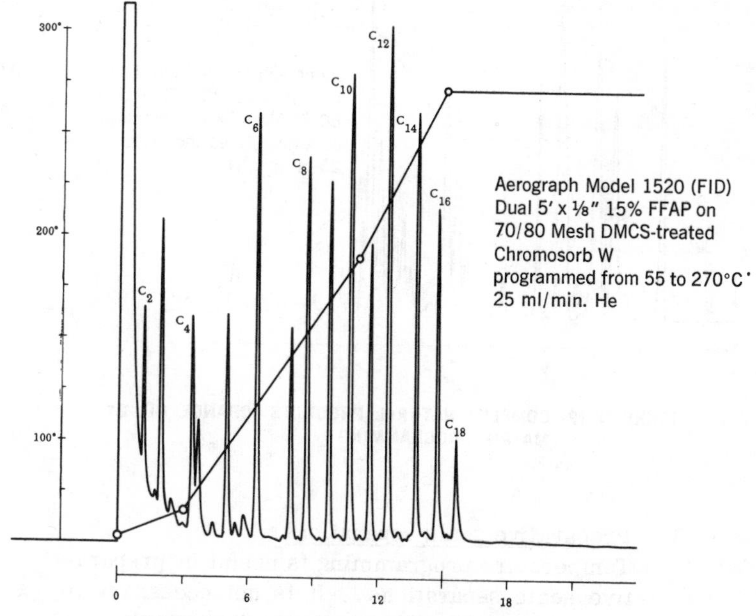

FIGURE IX-9 —WIDE BOILING RANGE SAMPLES—C_2 TO C_{18} ALCOHOLS ON FFAP

and dual detectors. The boiling points range from 70°C to 340°C. Analysis time is only 18 minutes. This type of analysis is possible only with temperature programming.

2. Complex natural products

Orange oil is typical of natural products. A low starting temperature is required to separate the volatile components. The higher boiling peaks, however, would require days to elute at a low isothermal temperature. Figure IX-10 shows the use of Matrix programming to stop programming, allow the two peaks at 35 minutes to resolve, and then continue at a new programming rate. The final temperature is determined by the liquid phase stability.

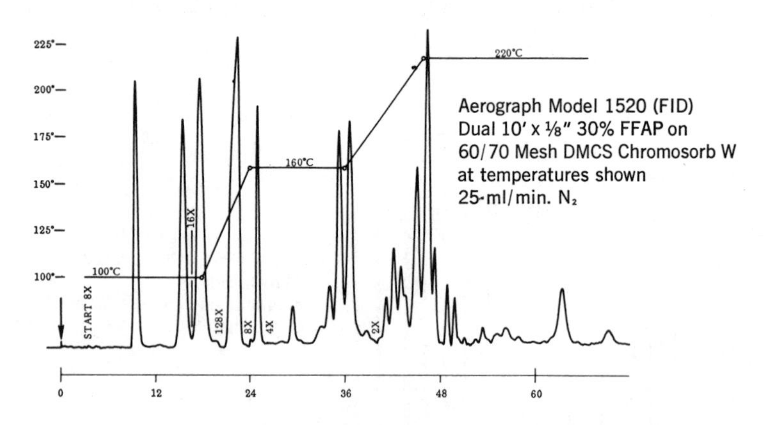

FIGURE IX-12—COMPLEX NATURAL PRODUCTS—ORANGE OIL BY MATRIX PROGRAMMING

3. Preparative scale separations

Temperature programming is useful in preparative scale separations. It is not necessary to instantaneously vaporize the entire sample. A slow injection does not effect resolution since at the low initial temperature the peaks are *frozen* at the front of the column. No loss of resolution will occur because the subsequent temperature programming will provide a good chromatogram. In addition, the more symmetrical and well-shaped peaks obtained in temperature programming allow for easier collection.

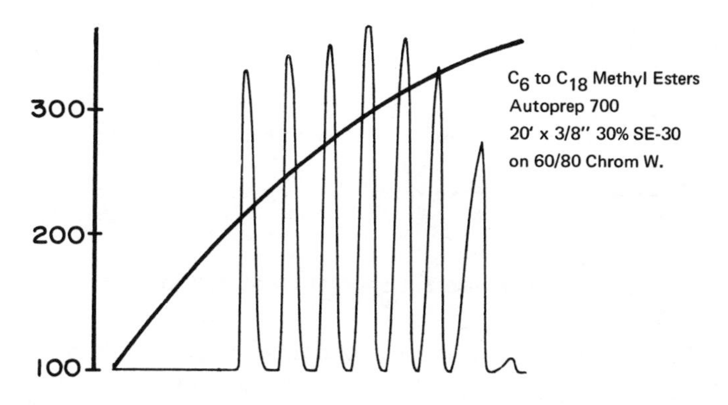

FIGURE IX-11—PREPARATIVE SCALE SEPARATIONS—7.0 ml FATTY ACID METHYL ESTERS

4. <u>Gas-solid chromatography</u>
   Adsorption columns display a stronger selectivity than partition liquids in many cases. Major disadvantages of G.S.C. however, are that analysis times are long and peak tailing makes quantitative analysis difficult.

   Both of these problems can be solved through temperature programming. As Figure IX-12 shows, it is possible to analyze, with one sample on one column, the gases oxygen, nitrogen, methane, and ethane. This type of analysis could be done with

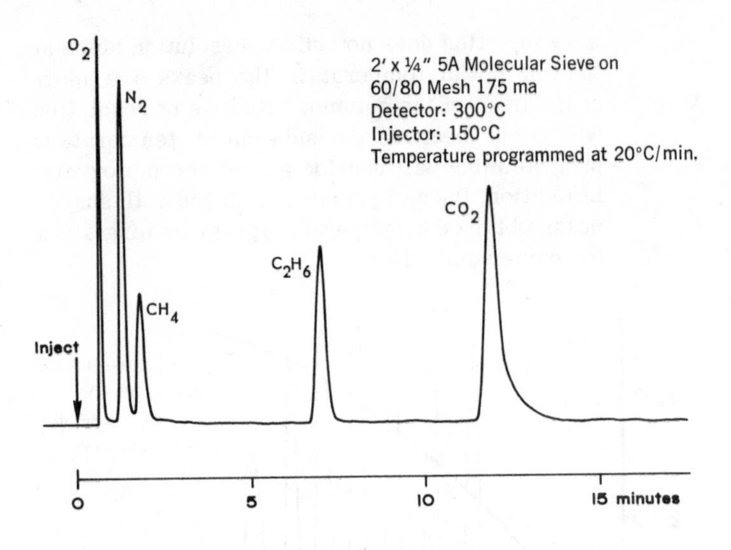

FIGURE IX-12—GAS SOLID CHROMATOGRAPHY BY PTGC

two columns and a column switching valve but it would not have the simplicity of temperature programming. The recovery of carbon dioxide is not considered to be quantitative, so it would be advisable to use another column arrangement if carbon dioxide is to be measured accurately.

5. Trace analysis

Trace analysis of high boiling components can be aided by PTGC. Repetitive injections of the sample can be made at low temperatures. The solvent is rapidly eluted. By keeping the column temperature low, the high boiling impurities are condensed on the column. When sufficient sample is present, the column can be programmed and the concentrated high boiling traces are eluted. In addition, the programming provides sharp peaks for higher boiling impurities. A good example of this is the on-column concentration step of odorants in natural gas.

F. QUALITATIVE ANALYSIS

Retention times are often used as an aid in peak identification. For retention times under programmed conditions to be useful for qualitative analysis, they must be reproducible. Most modern gas chromatographs provide precise reproduction of temperature programs and retention times.

A sample of C_7 - C_{22} normal paraffins, boiling range 98.4°C - 327°C, was used to illustrate the reproducibility of retention times. The mixture was analyzed on a polyphenylether capillary column as shown in Figure IX-13.

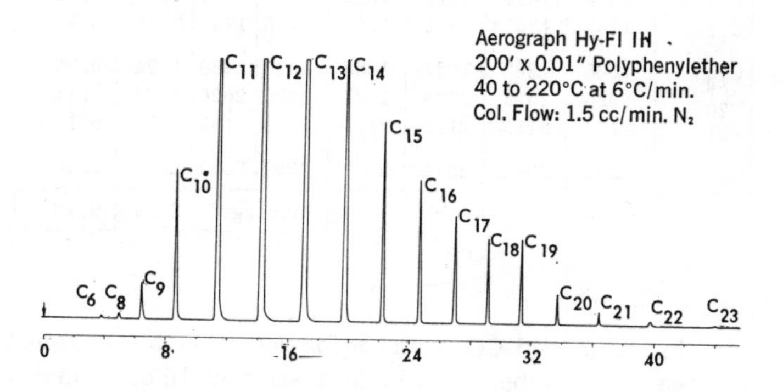

FIGURE IX-13—NORMAL PARAFFIN MIXTURE

The column oven was cooled to 35°C and then programmed at 6°C per minute. The sample was injected without interrupting the programmer as the oven temperature reached 40°C. Retention times were measured in seconds with an electronic digital integrator, Table IX-5 illustrates the reproducibility of retention times for several analyses.

TABLE IX-5—RETENTION TIME REPRODUCIBILITY

RETENTION TIME, SECONDS

COM-POUND	RUN I	RUN II	RUN III	RUN IV	RUN V	AVE.	σ ABS	σ REL. %
C_7	231	231	232	230	231	231	0.71	0.30
C_8	302	301	305	300	302	302	1.87	0.61
C_9	389	389	395	387	390	390	3.00	0.77
C_{10}	521	522	528	519	523	523	3.31	0.63
C_{11}	688	689	694	687	689	689	2.74	0.39
C_{12}	863	864	868	864	865	865	1.94	0.22
C_{13}	1032	1033	1037	1035	1034	1034	1.94	0.18
C_{14}	1191	1190	1195	1193	1192	1192	1.94	0.16
C_{15}	1339	1338	1342	1341	1341	1340	1.73	0.12
C_{16}	1483	1482	1486	1486	1485	1484	1.87	0.12
C_{17}	1623	1622	1626	1626	1626	1625	2.00	0.12
C_{18}	1752	1751	1755	1755	1755	1754	2.00	0.11
C_{19}	1885	1883	1887	1888	1886	1886	1.94	0.10
C_{20}	2025	2022	2028	2027	2026	2026	2.34	0.11
C_{21}	2187	2185	2192	2191	2189	2189	3.34	0.15
C_{22}	2388	2384	2394	2392	2390	2390	3.97	0.16
					Average		2.29	0.26

In isothermal GC, a plot of log retention times against carbon number results in a straight line. Purnell points out that this regularity is the result of a linear variation of ΔH_V the heat of vaporization with carbon number[5]. The entropy of solution shows a similar relationship.

In PTGC the plot of retention temperature is linear with carbon number or boiling point (see Figure IX-14 on next page). Baumann[6] reports a more nearly straight line when boiling point is used instead of carbon numbers. This technique aids greatly in the identification of peaks.

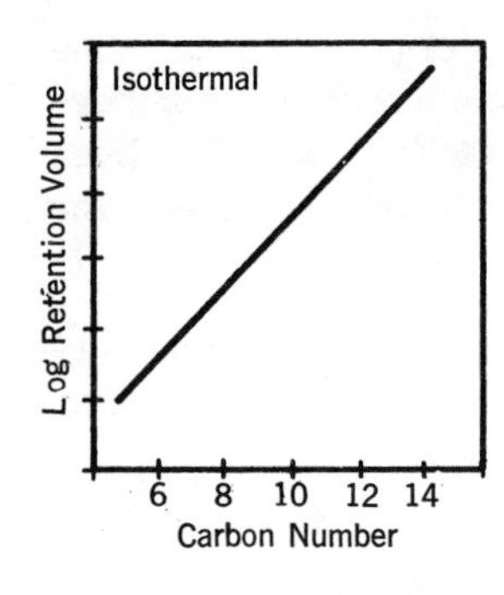

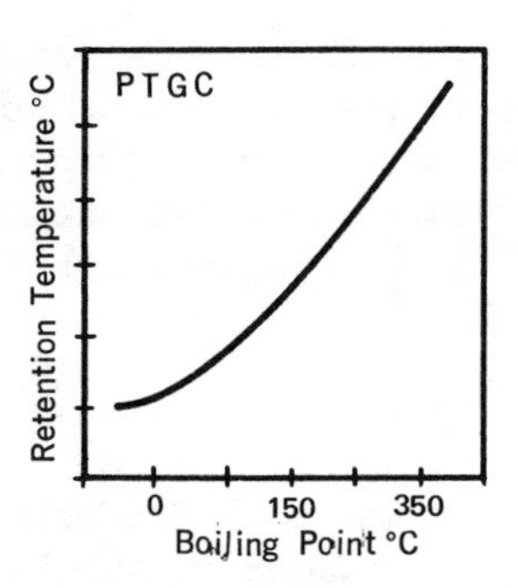

FIGURE IX-14—IDENTIFICATION BY RETENTION TEMPERATURE

G. QUANTITATIVE ANALYSIS

Quantitative analysis using programmed temperature operation has the same potential for precision and accuracy as isothermal operation, provided, of course that a precisely reproducible programmer is employed.

TABLE IX-5—REPRODUCIBILITY OF PEAK AREAS

NORMAL PARAFFINS

COMPOUND	RUN 1	RUN 2	COMPOUND	RUN 1	RUN 2
C_7	0.03	0.03	C_{15}	5.43	5.40
C_8	0.16	0.17	C_{16}	3.45	3.49
C_9	1.09	1.12	C_{17}	4.28	4.34
C_{10}	4.50	4.51	C_{18}	1.74	1.72
C_{11}	16.18	16.22	C_{19}	1.95	1.89
C_{12}	24.14	24.08	C_{20}	0.70	0.68
C_{13}	22.33	22.33	C_{21}	0.34	0.34
C_{14}	13.40	13.29	C_{22}	0.11	0.10

To illustrate the reproducibility of peak areas during PTGC, a sample of kerosene was analyzed. Temperature was programmed from 60°C to 230°C. Even though the baseline drifted due to column bleed, the areas were reproducible.(see Table IX-6).

A comparison of the accuracy obtainable during isothermal and programmed temperature operation was made by analyzing a mixture of C_{10} to C_{13} normal paraffins. The percentage compositions measured are shown in Table IX-7.

TABLE IX-7—COMPOSITION C_{10} TO C_{13} MIXTURE

COMPOUND	WEIGHED	ISOTHERMAL		TEMP. PROG. *	
n-C_{10}	11.66%	11.54	0.02%	11.66	0.06%
n-C_{11}	16.94%	16.91	0.02%	17.07	0.03%
n-C_{12}	33.14%	33.17	0.02%	33.15	0.03%
n-C_{13}	38.26%	38.38	0.03%	38.12	0.11%
TOTAL	100.00	100.00		100.00	

**Average and absolute standard deviation from 4 runs.*

BIBLIOGRAPHY

1. *Giddings, J.C., J. Chem. Educ., 39, 569 (1962).*
2. *Harris, W.E., Habgood, H.W., Programmed Temperature Gas Chromatography, John Wiley, New York, 1966.*
3. *Hawkes, S.J., and Mooney, E.F., Anal. Chem., 36, 1473 (1964).*
4. *Guild, L., et.al., Gas Chromatography, D. H. Desty, ed., p. 226, Butterworths, London 1958.*
5. *Purnell, H., Gas Chromatography, Wiley, New York 1962.*
6. *Baumann, F., et.al., Gas Chromatography, M. van Swaay, ed., Butterworths, London 1962.*

X. SERVICING OF INSTRUMENTS

A. INTRODUCTION

This section deals with proper maintenance and servicing of gas chromatographs. Although the specific examples and demonstrations used apply to Aerograph instruments, they can also apply to other gas chromatographs. Objectives of this chapter are:

1. Keep instrument "down time" to a minimum.
2. Avoid service problems by recognizing causes.
3. Ensure that your instrument is performing as well as it should.
4. Promote the idea of careful, systematic troubleshooting.

B. BASIC GUIDE FOR INSTRUMENT MAINTENANCE

1. Necessary tools and equipment

   A large investment in tools and test equipment is not necessary, but a few small hand tools are required. The following list is recommended.

   a. Tools

      (1) Screwdrivers:
      #1 x 3" Phillips
      #2 x 4" Phillips
      #2 Stubby Phillips
      1/16" x 6" Regular
      3/16" x 3" Regular
      1/4" x 4" Heavy duty regular

(2) Pliers:
Long nose
Common (gas pliers)
Small
Channel locking angle head pliers

(3) Wrenches:
3/8" x 7/16" Open end
1/2" x 9/16" Open end
6" Crescent
8" Crescent
12" Crescent for Cylinder connections
Allen wrenches (preferably long arm type)

.050"	1/8"
1/16"	9/64"
5/64"	5/32"
3/32"	3/16"
7/64"	

b. Test equipment

(1) Small volt-ohmmeter, 20,000 ohms per volt.

2. General troubleshooting

Before any repairs are started, READ THE INSTRUMENT INSTRUCTION MANUAL, particularly check servicing or troubleshooting sections. Also check operating sections. The "Chromatogram Interpretation" section of this book is a helpful place to start troubleshooting.

Before ever attempting to operate a gas chromatograph, it should be thoroughly checked for gas leaks. As shown in Chapter VIII, carrier gas leaks give rise to many problems. The carrier gas exit lines should be capped. The carrier gas input pressure is then increased to 60 psig. When the pressure has stabilized, turn off the main cylinder valve. If no leaks are present, the pressure should remain constant for 15 minutes. If the pressure does not hold, re-pressurize and check for leaks with dilute soap solution. Detailed leak checking and procedures are usually given in the instruction manuals.

3. The chromatograph recorder

Whenever problems arise that could also be caused by the recorder, it should be checked first for proper operation. Read the recorder instruction manual for more details. MAKE SURE RECORDER INPUT LEADS ARE PROPERLY CONNECTED (refer to recorder manual).

Verification of proper recorder operation may be quickly made by disconnecting the gas chromatograph from the recorder input, and connecting a short piece of wire (a paper clip works fine) between the + and - input connections. The recorder pen should go to electrical zero and remain there without any oscillation. At this point, the indicating pointer on the scale should indicate zero (mechanical zero) and the pen on the strip chart should indicate zero (chart zero), i.e., electrical, mechanical, and chart zero should all line up. The pen should not move with light finger pressure. Make any recorder adjustments that might be necessary. Some common problems are:

a. Sluggish pen movement

(1) GAIN adjustment too low, DAMPING adjustment overdamped.
(2) Check reference battery (where applicable).
(3) Check amplifier tubes.
(4) Check for low level AC voltage signal to input. Correct by installing $0.25\mu f$ (approximately) capacitor from either + or - connection to ground.

(5) Ensure the CHASSIS GROUND is a true EARTH GROUND. If possible, connect to water pipe.

(6) Check input leads for improper connection.

b. Jerky pen movement

(1) Dirty slide wire.
Clean the slide wire and contacts. Ethyl ether is a good solvent. Do not oil after cleaning.

(2) Mechanical binding

(3) Check tubes in amplifier

(4) Check for any severe line power changes (line transients, etc.) Isolate recorder and gas chromatograph, if necessary. Use stabilized transformer if necessary.

c. Pen oscillates

(1) GAIN adjustment too high. Adjust DAMPING control also.

(2) Poor ground connection.

(3) Input leads not properly connected.

d. Chart drive slips

(1) Tighten clutch. See recorder manual.

(2) Drive gear slipping on shaft. Tighten.

e. Attenuation not linear

(1) Poor ground connection.

(2) GAIN adjustment too low. Adjust DAMPING control also.

(3) Input leads not properly connected.

(4) Electrical and mechanical zero not at the same point and/or not at chart zero.

(5) Check amplifier tubes.

(6) Check reference battery (where applicable).

(7) SPAN ADJUST out of calibration (where applicable).

4. Troubleshooting temperature control and programming systems

The following problems are those most frequently encountered in heaters and temperature control systems:

a. No heat to detector (or column oven)

(1) Check fuses - Main Power, oven and controller (or programmer) fuses. Replace defective fuses.
(2) Open heater element - check with ohmmeter by disconnecting one heater lead. Set ohmmeter to highest scale reading and read element resistance. Infinite resistance indicates an open heater element - replace. Zero ohms indicates heater element is O.K.
(3) Check for loose wiring connections.
(4) Upper limit control switch set too low or defective.
(5) Defective temp. sensing probe. Check with ohmmeter. See instruction manual.
(6) Check tubes in detector oven controller (if applicable).

b. Full heat to detector (or column) oven regardless of position of control setting.

(1) Defective temp. sensing probe. Check with ohmmeter.
(2) Defective tubes in controller (when applicable)
(3) Defective silicone controlled rectifier (SCR) in controller (when applicable).

c. No heat to injectors

(1) Check fuses. Replace defective fuses.
(2) Open heater element. Check and, if necessary, replace, as described in "a (2)" above.
(3) Defective injector heater controller. See instruction manual.

d. Temperature in ovens not stable.

(1) Defective temperature sensing probe.
(2) Defective tubes in controller (when applicable)
(3) Gaps or holes in oven insulation. Crossover tubes should be capped when not in use.
(4) Defective pyrometer or loose pyrometer connections.

5. <u>Troubleshooting the thermal conductivity detector</u>

The following problems are those most frequently encountered in thermal conductivity systems.

a. Noisy baseline

(1) Carrier gas flow leaks. The most common places are column connections and septum.
(2) Contamination in columns or flow system, particularly injectors and any places in the flow system where cold spots might occur.
(3) Flow controller or cylinder regulator erratic.
(4) Loose wires or connections.
(5) Dirty switch contacts. Clean with spray cleaner or eraser stick (pencil).
(6) Defective detector.
(7) Defective power supply.
(8) Defective recorder.

b. Drifting baseline

(1) Carrier gas flow leaks. This is by far the most common cause of both drifting and noise.
(2) Contamination in column. Recondition
(3) Flow controller or cylinder regulator defective.
(4) Detector oven temperature not at equilibrium temperature. This could be due to insufficient time being allowed for oven to reach a stable temperature. Also see comments in "4 d".
(5) Contaminated detector. Clean detector cell (see instruction manual).
(6) Poor carrier gas flow through either detector or reference side of detector.

c. Recorder will not zero

(1) No carrier gas flow through reference side of detector
(2) Detector cell filaments out of balance or burned out.
(3) Recorder not connected and/or adjusted properly.
(4) Poor connection in the D. C. circuit of the instrument or interconnecting cable to recorder.

d. No filament current

(1) Blown fuse in D. C. power supply. Check and replace if defective.
(2) Defective power supply or power supply switch causing no output from supply.
(3) Burned out filaments. Can be isolated by checking schematic diagram in instruction manual. Test with ohmmeter.
(4) Defective current meter on instrument panel.

(5) Broken electrical connection in D. C. circuit of instrument.

e. Attenuator not linear

(1) Check resistors on attenuator for value and/or solder connections.
(2) Recorder electrical zero not at recorder scale zero.
(3) Poor connection between instrument and recorder, and/or recorder gain; damping, etc. out of adjustment. See section "3e".

f. Poor chromatogram presentation on recorder chart. Refer to Chapter VIII.

6. <u>Troubleshooting ionization detector systems</u>

The portions of the instrument ahead (upstream) of the detector are virtually identical to those used with thermal conductivity detectors, except that no reference gas is needed. Electrical systems used to control heaters, etc., are also the same, so are subject to the same troubleshooting methods outlined previously.

An electrometer acts as an amplifier for the detector signal, and supplies an output compatible with the recorder. Therefore, the input must be matched to the detector and the output matched to the recorder.

If there is difficulty with a detector system using an electrometer, it is desirable to isolate the problem to a particular component within the system (detector, electrometer, or recorder).

a. Isolation of the component causing the problem

(1) Check recorder as outlined in section "3" of this section. If the recorder operates

normally, reconnect it to the system.

(2) Check electrometer by disconnecting signal input cable (leading from the detector) and shielding the input to the electrometer with a piece of aluminum foil. Be careful that the foil only shields the input and does not short it to chassis ground. The foil serves to insure that no stray signals will be picked up, and amplified by the electrometer.

If the operation agrees with specifications outlined in the instruction manuals, the electrometer is not at fault and the problem is being introduced into the system by the detector.

If at this point the electrometer is found to be at fault, the problem is usually poor stability (drifting or noisy baseline), or a problem with balancing the baseline at recorder zero.

(a) Drifting baseline

Check the electrometer tubes and replace the defective ones. Allow several hours for electrometer to restabilize before rechecking.

(b) Noisy baseline

1. Check for any defective tubes as above.
2. Check for bad connection between electrometer and recorder.
3. Check for any poor or dirty switch contacts in the electrometer.

(c) Cannot balance electrometer at recorder zero.

1. Check electrometer the same as for noisy baseline.
2. Check instruction manual and rebalance electrometer input circuit, if applicable.

If operation of the recorder and electrometer are found to be normal, check detector operation as outlined in the instruction manual.

Some typical troubles with specific detectors are as follows:

a. Flame ionization detector

(1) **Flame will not stay lit.**

(a) Flame burner tip restricted. Clean and check orifice size.
(b) Insufficient hydrogen. Check for any leaks and restrictions in supply. Use more hydrogen.
(c) Insufficient air. Same as (b) above. Use more air.
(d) Improper flow rates. Check the air, carrier gas, and hydrogen rates and adjust to specified optimum flows.
(e) Too large a sample injected into instrument.

(2) Noisy baseline

(a) Inspect the detector for cleanliness. Dirt on the detector parts (especially electrical connectors) can cause the noise. Dirt can be caused by sample residue or liquid phase residue and is best removed by immersing the complete detector in an ultrasonic cleaner.
(b) Noise can originate from particles or residue in the flame base passage-

ways. This is an often overlooked point. Clean the base just as you would the detector.

(c) Excessive column bleed from poor columns or excessive temperature can often be identified by viewing the flame. A clean burning flame is not visible. A dirty flame burns orange or bluish.

(d) Dirty carrier gas, air or hydrogen can cause noise and should be investigated if normal checks do not eliminate noise.

(e) Improper balance of air, hydrogen, and carrier gas to the detector is a common cause of noise. Check instruction manual for optimum flow rates.

(3) Drifting baseline

(a) Drifting baseline is usually caused by column bleed or faulty flow system (leaks etc.).

b. Electron capture detector

The following are some typical troubles encountered when using an electron capture detector.

(1) Unstable baseline

(a) Carrier gas leaks. Check for leaks as with any other system.

(b) Column bleeding. Disconnect column from detector and condition column at approximately 50°C higher than normal operating temperature.

(c) Oven heat cycling. Check as outlined in section "4" of this chapter.

(d) Carrier gas filter is saturated with contaminants. Regenerate filter as outlined in instruction manual.

(e) Air back diffusing into the detector.

(f) Faulty pressure regulator on carrier gas cylinder.

(g) Column oven cover is not properly closed.

(h) Poor earth grounding of complete instrument system.

(2) Low standing current

(a) Improper connections to the detector.

(b) Low carrier gas flow. For optimum flow through detector check instruction manual.

(c) Tritium foil reversed in cell.

(d) Tritium foil contaminated by sample or column bleed. Clean as outlined in instruction manual.

(e) Carrier gas leaks. Check as with any other system.

XI. SPECIAL TOPICS

CAPILLARY COLUMNS[1,2]

Capillary columns were developed by M. J. E. Golay in 1956, in the course of theoretical considerations of packed column behavior. Capillaries are long, open tubes of small diameter. They have high efficiencies, low sample capacity and low pressure drop. Capillary or open tubular columns range from 0.01 to 0.03 inches in I.D. and from 100 to 500 feet in length, though longer ones have been made. The inside wall of the tubing is coated with a thin film of liquid phase. Naturally, the sample size permitted with such a small diameter column is small, and requires the use of high sensitivity detectors such as flame ionization, or argon β -ray ionization.

Sample sizes for capillary columns are normally 100 to 1000 fold smaller than packed columns. Since it is not practical to inject 0.001 - 0.01 μl samples directly an inlet splitter device is required. A typical inlet splitter is shown in Figure XI-1.

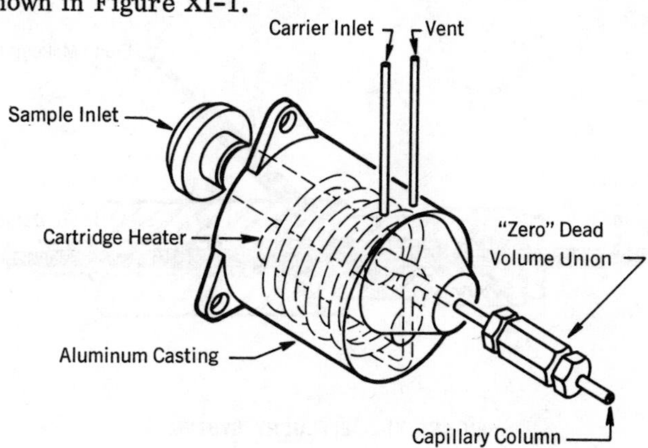

FIGURE XI-1—INLET SPLITTER

The carrier gas, usually helium or nitrogen, is introduced at the carrier inlet. The carrier mixes with the sample injected and is split into two streams - one going to the capillary column and the other being vented to the atmosphere by means of an adjustable vent. Usual practice is to inject samples on the order of one microliter and split the sample about 100:1. In order to reduce band broadening due to void spaces, "zero" dead volume connections are used to attach the column to the inlet-splitter and detector.

The entire flow system of a capillary column instrument places a stringent requirement on the detection system. The detector must have high sensitivity because of the small sample, must be capable of fast response because of narrow peaks, and must have low volume to avoid band broadening.

To eliminate or minimize band broadening in the effluent system leading to the detector, a good practice is to add a makeup gas to increase the linear velocity and decrease the residence time of the components as they are swept into the detector. Figure XI-2 illustrates an effluent system especially designed for capillary columns. If the capillary column employs a flame ionization detector then hydrogen gas serves the dual purpose of fuel and make-up gas.

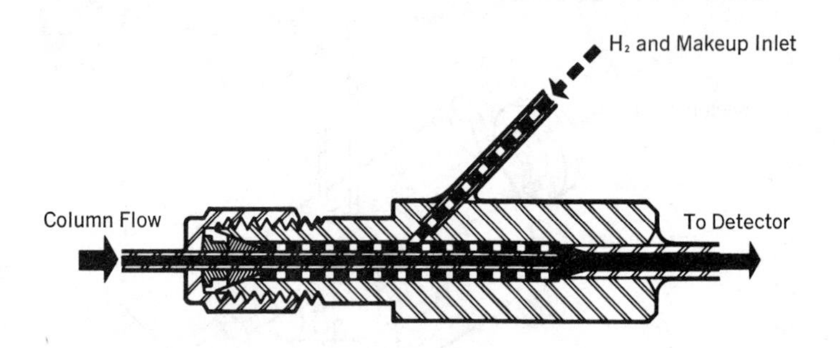

FIGURE XI-2—EFFLUENT SYSTEM

The make-up gas also allows one to optimize the operating conditions of the flame ionization detector. Optimum flow rates are in the ratio of 10:1:1 for air, hydrogen, and carrier gas respectively; thus the addition of 25 to 30 ml per minute of make-up gas to the 1 to 5 ml/minute capillary column flow rate optimizes the flame ionization conditions.

Applications

Capillary columns have the high efficiencies required for the separation of closely spaced peaks. Very good resolution can frequently be obtained in a reasonable time. Therefore, capillaries are recommended for the separation of complex mixtures, and the resolution of closely related isomers. Figure XI-3 is a typical capillary chromatogram. The crude distillation cut was separated on a column designed by Schwartz and Brasseaux [3].

The column is a 200' x 0.01" I.D. capillary, coated with hexadecane, hexadecene and Kel-F oil (HHK). The temperature is 28°C. The column separates all saturated C_4-C_7 hydrocarbons and resolves the C_8 hydrocarbons quite well. The 45 C_8's are separated into 36 peaks. This column has found extensive use in the petroleum industry to analyze crude oil cuts, gasoline, reformate, platformate, and various other refinery samples (see next page).

Figure XI-4 shows the separation of saturated and unsaturated C_6 - C_9 aromatic compounds. The separation is carried out at 85°C on a 200' x 0.01" I.D. column, coated with 6-ring polyphenylether. All possible C_6 - C_9 aromatics are shown except for 1, 2, 4 trimethylbenzene and paramethylstyrene. Ucon LB550X also gives a very good

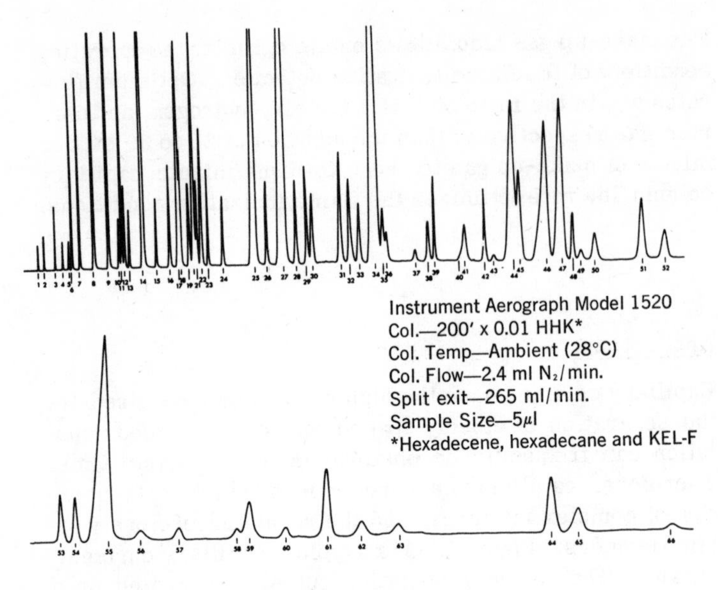

Retention time relative
to n-Heptane (27.0 minutes)
are shown.

1.	Isopentane	0.152
2.	*n*-Pentane	0.175
3.	2,2-Dimethylbutane	0.219
4.	Cyclopentane	0.249
5.	2,3-Dimethylbutane	0.272
6.	2-Methylpentane	0.284
7.	3-Methylpentane	0.314
8.	*n*-Hexane	0.370
9.	Methylcyclopentane	0.426
10.	2,2-Dimethylpentane	0.464
11.	Benzene	0.476
12.	2,4-Dimethylpentane	0.487
13.	2,2,3-Trimethylbutane	0.506
14.	Cyclohexane	0.562
15.	3,3-Dimethylpentane	0.610
16.	1,1-Dimethylcyclopentane	0.663
17.	2-Methylhexane	0.701
18.	2,3-Dimethylpentane	0.707
19.	1, *cis*-3-Dimethcyclopentane	0.740
20.	3-Methylhexane	0.761
21.	1 *trans*-3-Dimethcyclopentane	0.771
22.	1 *t*-2-Dimethcyclopentane	0.789
23.	3-Ethylpentane	0.815
24.	2,2,4-Trimethpentane	0.828
25.	*n*-Heptane	1.000
26.	1, c-2-Dimethcyclopentane	1.044
27.	Methylcyclohexane	1.116
28.	1, 1, 3-Trimethylcyclopentane 2,2,3,3-Tetramethylbutane	1.158
29.	2,2-Dimethylhexane	1.202
30.	Ethylcyclopentane	1.222
31.	2,5-Dimethylhexane	1.330
32.	2,4-Dimethylhexane 2,2,3-Trimethylpentane	1.368
33.	1, *t*-2, c-4-Trimethcyclopentane	1.403
34.	Toluene	1.475
35.	1, *t*-2,c-3-Trimethcyclopentane	1.496
36.	3,3-Dimethylhexane	1.510
37.	2,3,4-Trimethylpentane	1.623
38.	1,1,2-Trimethcyclopentane	1.672
39.	2,3,3-Trimethpentane	1.695
40.	2-Methyl-3-ethylpentane	1.812
41.	2,3-Dimethhexane	1.812
42.	1, c-2, *t*-4-Trimethcyclopentane	1.898
43.	1, c-2, *t*-3-Trimethcyclopentane	1.930
44.	3,4-Dimethylhexane 2-Methylheptane 4-Methylheptane	2.005
45.	3-Meth-3-ethpentané 1, c-2, c-4-Trimethcyclopentane	2.027
46.	3-Ethylhexane 3-Methylheptane	2.132
47.	1,1-Dimethcyclohexane 1, *t*-4-Dimethcyclohexane	2.193
48.	1, c-3-Dimethcyclohexane	2.240
49.	1-Meth-*t*-3-ethcyclopentane	2.266
50.	1-Meth-*t*-2-ethcyclopentane 1-Meth-c-3-ethcyclopentane 1-Meth-1-ethcyclopentane	2.319
51.	1, c-2, c-3-Trimethcyclopentane	2.503
52.	1, *t*-2-Dimethcyclohexane	2.591
53.	1, c-4-Dimethcyclohexane	2.796
54.	1, *t*-3-Dimethcyclohexane	2.845
55.	*n*-Octane	2.979
56.	Isopropylcyclopentane	3.036
57.	1-Meth-c-2-etncyclopentane	3.104
58.	*n*-Propylcyclopentane	3.250
59.	Ethylcyclohexane	3.525
60.	C_9 Paraffin	3.672
61.	Ethylbenzene	3.828
62.	C_9 Paraffin	3.967
63.	C_9 Paraffin	4.112
64.	*p*-Xylene	4.705
65.	*m*-Xylenc	4.804
66.	C_9 Paraffin	5.470

FIGURE XI-3—CAPILLARY COLUMN—CRUDE DISTILLATION CUT

separation of most C_6 - C_{10} aromatics including many of the lower boiling C_{11}'s.

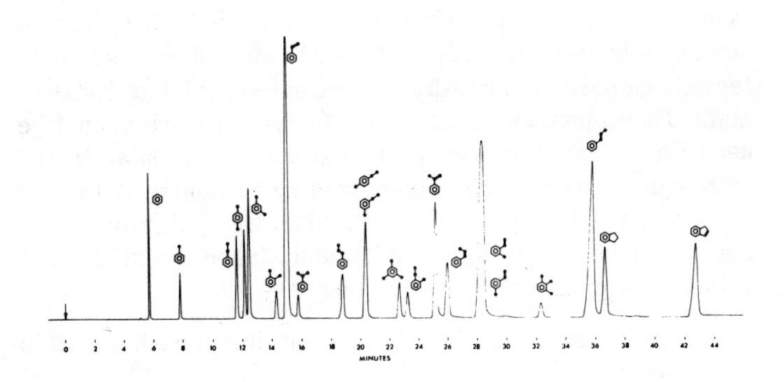

FIGURE XI-4—C_6-C_{11} AROMATICS

PREPARATIVE SCALE GAS CHROMATOGRAPHY

In gas chromatography, the chemist must always choose between resolution, speed and sample capacity. In some cases a workable compromise between two of these factors can be obtained, but always at the loss of the third.

In most cases, the goal of preparative scale separation is not sample capacity, but is the throughput, or grams per hour, which can be obtained. A particularly difficult separation may demand a long retention time and a small sample charge to get the desired resolution. However, if the system can be automated, these difficulties may be allowed and usable amounts of purified material may be obtained.

Scaling up[4]

Any column can be used to prepare a sample. Of course, the amount of sample which can be handled on an 1/8" column would be small. The most obvious way to increase sample size would be to increase the column diameter. There are many equally important considerations involved in scaling up to increase throughput.

First, the best possible *analytical column* for the separation of all components should be determined. The bleed rate of the liquid phase should be considered. It is best to select a phase with the lowest bleed rate, otherwise, collected samples will be badly contaminated with liquid phase. Many liquid phases suitable for analytical work cannot be used for preparative work, although a sample contaminated with liquid phase may be cleaned up by rerunning it through a short, low-loaded column of a different polarity. Two low-bleed liquid phases, SE-30 and Carbowax 20M, will solve most preparative problems.

Preparative runs should be made at the lowest feasible *temperature*.

One general rule is that the column should be 200°C. below the boiling point of the liquid phase (or vapor pressure is less than 1 mm at operating temperature) to minimize bleeding. If the temperature must be raised, then the percentage liquid load should be lowered; the column length can sometimes be increased to maintain resolution.

The *support* should be considered. Chromosorb P gives the most efficient columns (highest number of plates) but has been shown to have a deleterious effect on oxygenated compounds. Shorter retention times are obtained using Chromosorb W. If possible, Chromosorb T should be used for particularly polar or reactive compounds, as it is the most inert support.* It has been found that a narrow range of mesh size is best. Under 150°C. a 60/80 mesh size has been found optimum, over 150°C. - 45/60 mesh.

SE-30 columns should be prepared using Chromosorb W. It has been found that SE-30 on Chromosorb P bleeds more, even after conditioning, than will SE-30 on Chromosorb W. Taking all these factors into consideration, Chromosorb W is the best support for general preparative work.

**Chromosorb A has been developed as a preparative support and it is intermediate between Chromosorb P & W with respect to inertness and surface area.*

The usable sample size on a 3/8" column is roughly ten times the maximum sample usable on an 1/8" column and 3.5 to 4 times that for a 1/4" column. Doubling the length of a column also approximately doubles the capacity. There are many exceptions. These amounts are approximations. Preparative columns are usually overloaded, as peak symmetry is unimportant. The more separation between peaks, the larger the usable sample.

Column systems

Two approaches of current interest are:

(1) To use long columns with a small increase in diameter
(2) To use extremely wide diameter columns of short length

Recent improvements in support materials have allowed extremely long columns to be constructed with a reasonable pressure drop up to 250 feet long, outside diameter 3/8 inches, packed with 20/30 mesh Chromosorb A. The total pressure drop was approximately two atmospheres and a flow rate of 150 ml/min was obtained. The use of long, thin columns has the advantage of high resolution, adequate sample throughput and low operating costs. The one factor which must be sacrificed is time. Even this factor can be improved by the use of multiple injections, or flow rates faster than normally employed since at preparative scale, resolution is dependent on sample size rather than flow rate.

Figure XI-5 shows a repetitive separation of alpha and beta methyl naphthalene. When the chemist is faced with the problem of separating a difficult two or three component mixture, he is forced to use a long column to gain adequate resolution, which results in a long retention time. By operating isothermally and injecting the sample at proper time intervals (0-7 in Figure XI-5) throughput is increased.

The use of large diameter (4 inch I.D.) columns is also interesting. These columns have the advantage of sample

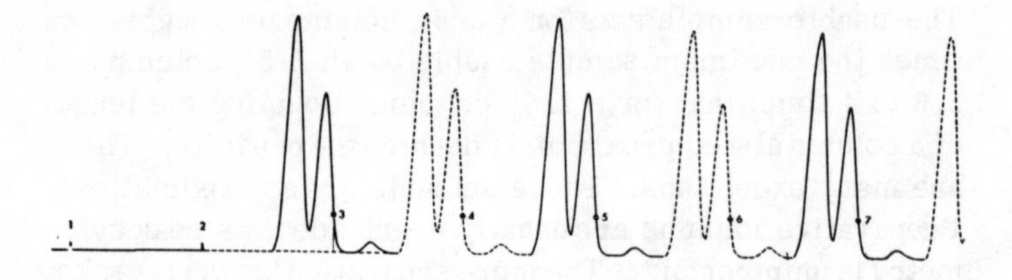

FIGURE XI-5—MULTIPLE INJECTION OF ALPHA AND BETA METHYL NAPHTHALENE

capacity and speed. In point of fact, it can be stated that liter per hour separations are possible. Unfortunately, the separating efficiency is limited to about 1,000 theoretical plates and the operating costs (columns, carrier gas, etc.) could be prohibitive in some cases.

Applications

Figure XI-6 demonstrates the separation of 0.5 ml of C_6 isomeric hydrocarbons using a 20' x 3/8" column of 30% tricresyl phosphate on 60/80 Chromosorb W. The column was operated at 65°C.

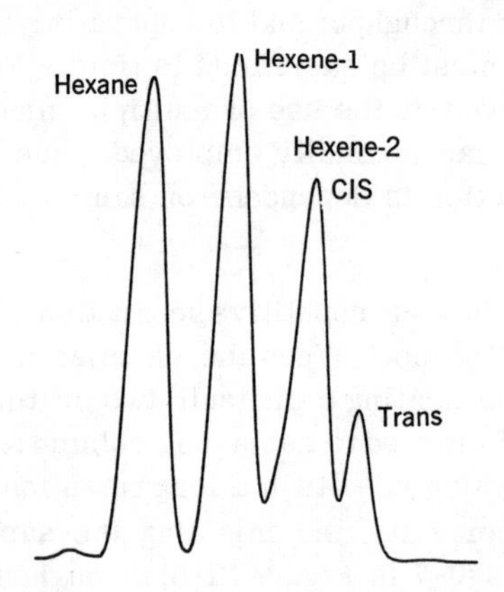

FIGURE XI-6—C_6 ISOMERIC HYDROCARBONS

Many samples cannot be resolved in a single pass through the chromatograph. The preparative separation of methyl-styrene dimers is a typical example. Figure XI-7 shows a 50 ml injection to resolve the monomers, dimers, and trimers.

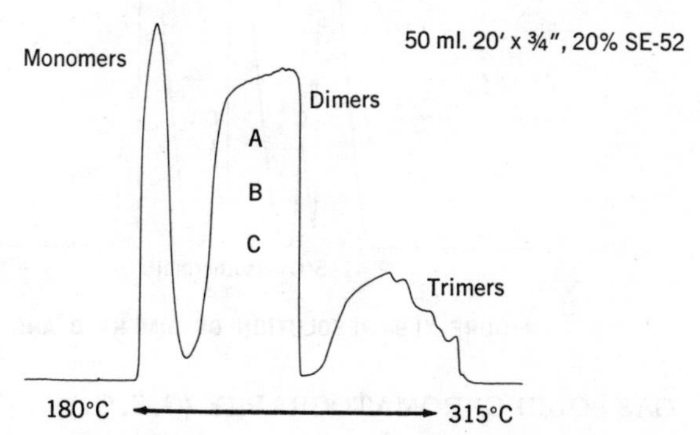

FIGURE XI-7—RESOLUTION OF METHYL-STYRENE, MONOMERS, DIMERS, AND TRIMERS

The collected styrene dimers are rerun as shown in Figure XI-8. Dimer A is collected with high purity; however, dimers B and C must be chromatographed a third time to obtain the desired purity (Figure XI-9).

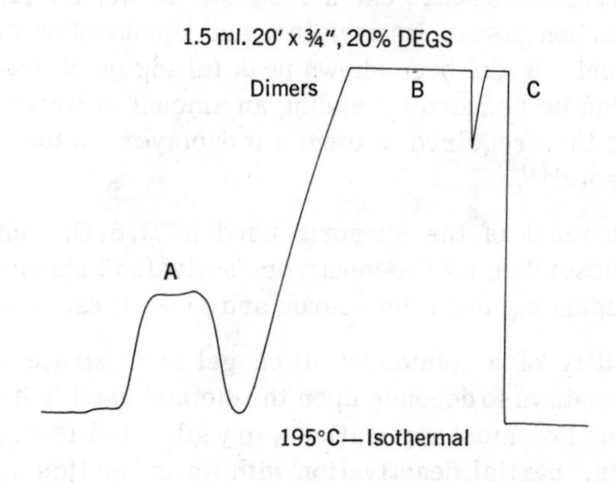

FIGURE XI-8—RESOLUTION OF STYRENE DIMERS

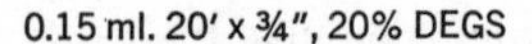

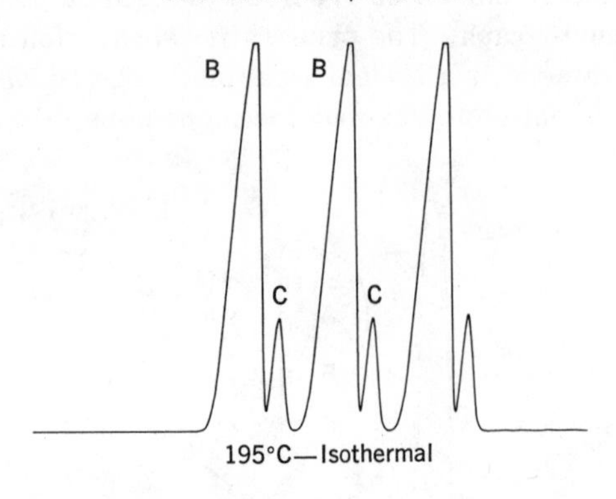

FIGURE XI-9—RESOLUTION OF DIMERS B AND C

GAS SOLID CHROMATOGRAPHY (G.S.C.)

G.S.C. represents a special part of chromatography where samples, mostly permanent gases and low molecular weight hydrocarbons, are separated by adsorption on an active solid rather than a liquid phase. Some of these active solids are alumina, charcoal, silica gel and various molecular sieves.

Alumina has been used extensively for the separation of the hydrocarbon gases. Its usefulness is dependent on its prior treatment. It has been shown peak tailing on alumina columns can be reduced by adding an amount of water equivalent to that required to form a monolayer on the surface of the solid[5].

As with most of the supports used in G.S.C., charcoal (often described as "coconut" or "activated" charcoal) can vary depending upon the source and prior treatment.

The ability of a column of silica gel to separate any two components also depends upon the method used in its preparation. For most applications, dry silica gel is employed; however, partial deactivation with water or liquid phases has been employed in specific analyses.

Molecular sieves are synthetic zeolites which are widely used as adsorbents in gas chromatography. These compounds are capable of separating materials based on their molecular size and configuration, and they adsorb molecules that have polar or polarizable properties. The latter characteristic enables the preferential adsorption of polar, unsaturated and aromatic compounds.

The molecular sieves employed in gas chromatography are 4A, 5A, 10X and 13X. The structure that is representative of one of these crystals (4A) is:

$$Na_{12}\ (AlO_2)_{12}\ (SiO_2)_{12}\ \cdot 27H_2O$$

Figure XI-10 compares the separating properties of 5A and 13X on an oxygen-nitrogen mixture.

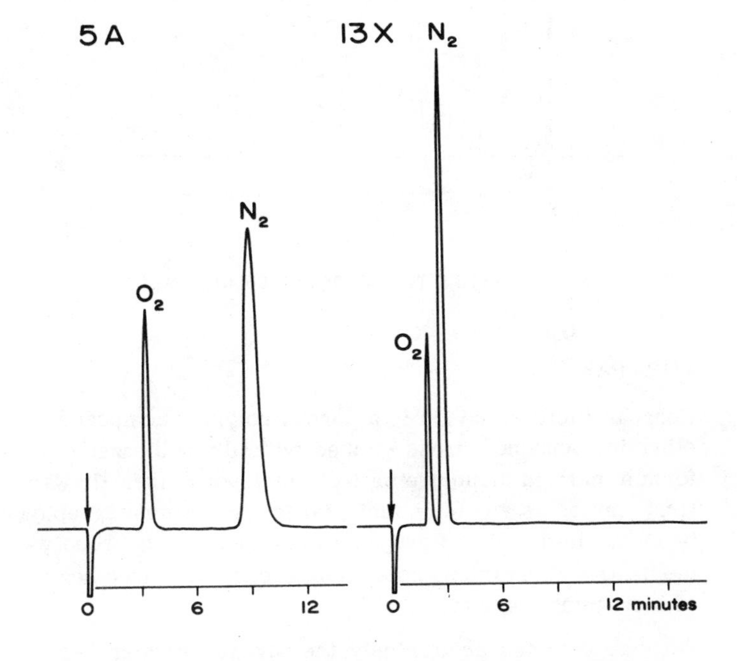

FIGURE XI-10—SEPARATION OF OXYGEN-NITROGEN MIXTURE ON MOLECULAR SIEVE 5A AND 13X

Figure XI-11 shows a typical example of a series column system in the analysis of a mixture of gases. The first column, a 20 inch silica gel is installed in the column oven and held at 100°C. The second column, a 20 foot, 10% Molecular Sieve 5A - 90% Molecular Sieve 13X is connected externally from the sensing side of the T.C. detector to the reference side of the detector. This system permits complete analysis for a single sample injection.

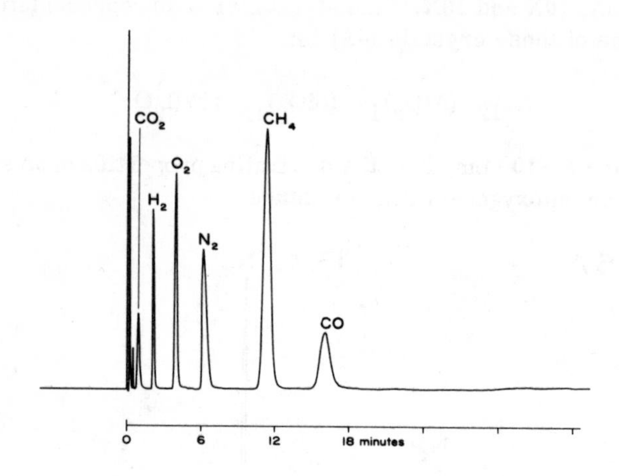

FIGURE XI-11—A TYPICAL SERIES COLUMN SYSTEM

PORAPAK®*

Porapak (Hollis type 19) is a porous polymer composed of ethylvinylbenzene cross-linked with divinylbenzene to form a uniform structure of a distinct pore size. It was first applied as a column material to gas chromatography by O. L. Hollis [6]. Porapak can be modified by copolymerization with various polar monomers to produce beads of increasing polarity.

Porapak columns contain only the porous polymer beads screened to a definite mesh range - no liquid phase or solid

**Porapak is a registered trademark of Waters Associates.*

support is used. The porous polymer beads serve the function of both the liquid phase and the solid support. Apparently the solute molecule partition directly from the gas phase into the amorphous polymer.

There are six Porapak materials available, type P, Q, R, S, T and N. Types P and Q are non polar Phases; however, type P has a much larger average pore size than type Q. Types R, S, T and N are moderately polar beads which have been modified with specific monomers. Porapak N has been developed for separation of formaldehyde mixtures and for the separation of acetylene and ethylene. All beads are available in various particle sizes ranging from 50/80 mesh to 150/200 mesh. All the polymers are stable to 250°C. except Porapak T (200°C).

The nature of the porous polymer largely determines the properties of the columns. One of the remarkable properties of these columns is the rapid elution of water and other highly polar molecules with little or no tailing.

Retention data is very constant since the Porapak columns usually have no liquid phase to be lost by continual bleeding. Since Porapak does not bleed, these columns can be employed with the ultra sensitive helium detector and temperature programmed with the flame ionization instrument.

Porapak is used primarily for efficient separation of a wide variety of relatively low molecular weight materials. The type Q material is the most versatile for the majority of separations. A study[7] of optimum conditions for 1/8" O.D. Porapak Q column has shown that the best results for most polar and non-polar materials are obtained using a mesh size of 100-120 with columns under 10 feet in length. Optimum flow rate is from 25 to 30 cc per minute. Sample size should be kept under 10 μl for mixtures with each individual component less than 0.03 μl.

Porapak columns should be conditioned prior to use to remove residual solvents from the polymerization process.

Higher efficiency columns will be obtained if Porapak is preconditioned before filling a column. This is done by placing the material in a large tube, at a temperature of 230°C for 18 to 20 hours. A flow rate of 50 ml/min (N_2 or He) should be maintained. Preconditioning is also recommended if the Porapak is to be partially deactivated by coating with a liquid phase.

Molecular weight should be less than 300 for components separated in a Porapak column. Porapak Q is especially useful for baseline separation of aqueous solution (Figure XI-12 & 13), low molecular weight hydrocarbons (Figure XI-14), alcohols (Figure XI-15), esters (Figure XI-16), ketones (Figures XI-17) and low molecular weight materials containing halogen and sulfur (Figure XI-18).

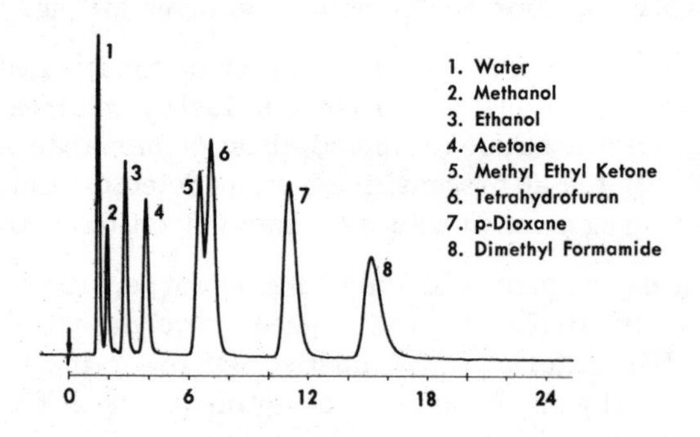

Aerograph A-90-P, 1.0 μl, 6' x ¼" OD, SS, Porapak Q, 150-200 mesh, 220°C., He 37 cc/min., TC.

FIGURE XI-12—WATER IN SOLVENTS

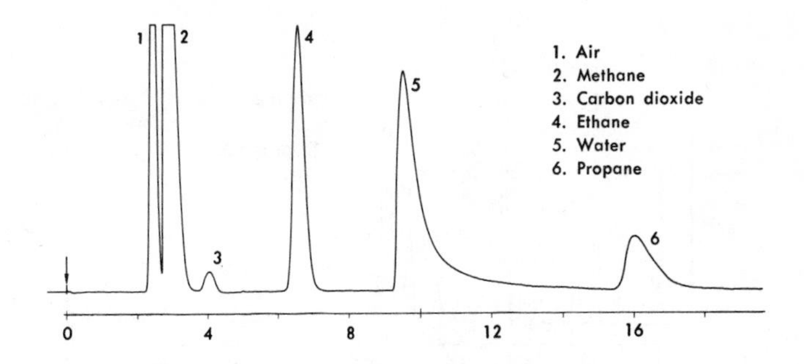

Aerograph A-90-P, 10′ x ⅛″ OD SS, Porapak Q, 150-200 mesh, 104°C, He 80 cc/min.

FIGURE XI-13—WATER AND HYDROCARBONS

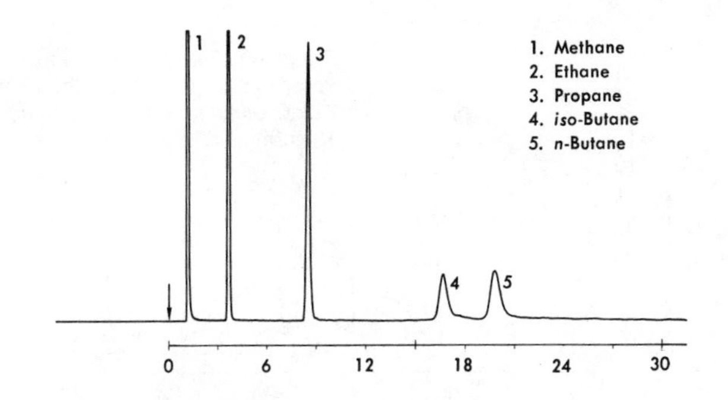

Aerograph 204, 50 µl, 7′ x ⅛″ OD SS, Porapak Q, 150-200 mesh, 50°C isothermally for first two minutes, then programmed 50°C-100°C @ 15°C/min, He 60 cc/min, FID.

FIGURE XI-14—NATURAL GAS

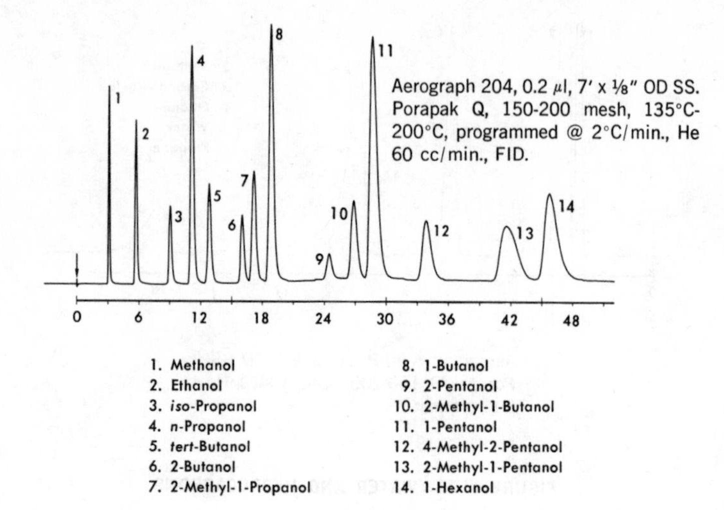
Aerograph 204, 0.2 μl, 7′ x ⅛″ OD SS. Porapak Q, 150-200 mesh, 135°C-200°C, programmed @ 2°C/min., He 60 cc/min., FID.
0 6 12 18 24 30 36 42 48
1. Methanol
2. Ethanol
3. *iso*-Propanol
4. *n*-Propanol
5. *tert*-Butanol
6. 2-Butanol
7. 2-Methyl-1-Propanol
8. 1-Butanol
9. 2-Pentanol
10. 2-Methyl-1-Butanol
11. 1-Pentanol
12. 4-Methyl-2-Pentanol
13. 2-Methyl-1-Pentanol
14. 1-Hexanol

FIGURE XI-15— C_1-C_6 ALCOHOLS

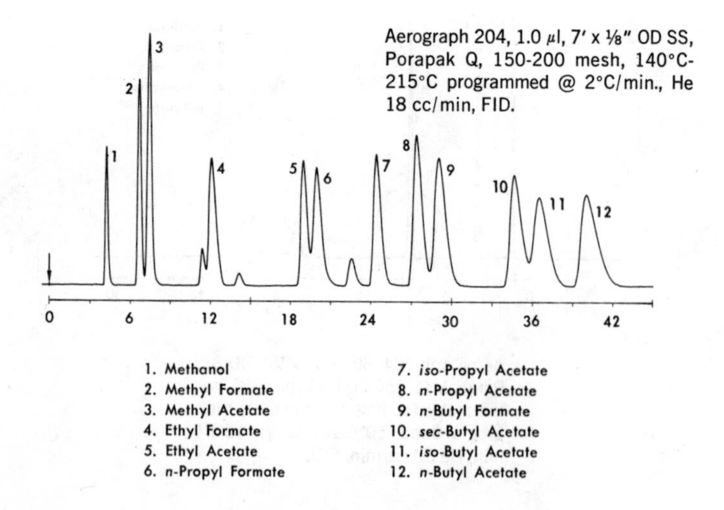
Aerograph 204, 1.0 μl, 7′ x ⅛″ OD SS, Porapak Q, 150-200 mesh, 140°C-215°C programmed @ 2°C/min., He 18 cc/min, FID.
0 6 12 18 24 30 36 42
1. Methanol
2. Methyl Formate
3. Methyl Acetate
4. Ethyl Formate
5. Ethyl Acetate
6. *n*-Propyl Formate
7. *iso*-Propyl Acetate
8. *n*-Propyl Acetate
9. *n*-Butyl Formate
10. *sec*-Butyl Acetate
11. *iso*-Butyl Acetate
12. *n*-Butyl Acetate

FIGURE XI-16—ESTERS

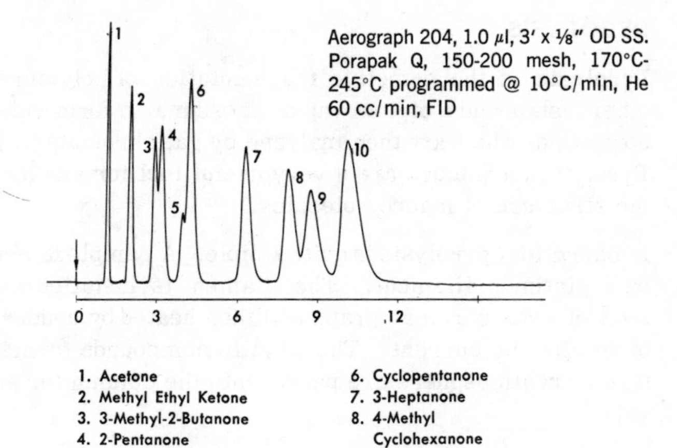
Aerograph 204, 1.0 µl, 3′ x ⅛″ OD SS. Porapak Q, 150-200 mesh, 170°C-245°C programmed @ 10°C/min, He 60 cc/min, FID
1
2
3
4
5
6
7
8
9
10
0
3
6
9
12
1. Acetone
2. Methyl Ethyl Ketone
3. 3-Methyl-2-Butanone
4. 2-Pentanone
5. 3,3-Dimethyl-2-Butanone
6. Cyclopentanone
7. 3-Heptanone
8. 4-Methyl Cyclohexanone
9. 2-Octanone
10. Acetophenone

FIGURE XI-17—KETONES

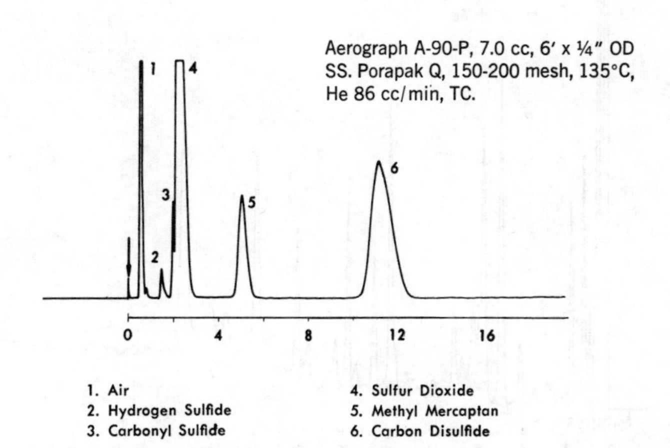
Aerograph A-90-P, 7.0 cc, 6′ x ¼″ OD SS. Porapak Q, 150-200 mesh, 135°C, He 86 cc/min, TC.
1
2
3
4
5
6
0
4
8
12
16
1. Air
2. Hydrogen Sulfide
3. Carbonyl Sulfide
4. Sulfur Dioxide
5. Methyl Mercaptan
6. Carbon Disulfide

FIGURE XI-18—SULFUR COMPOUNDS

PYROLYSIS

Pyrolysis is the selective fragmentation of polymers or other compounds of low vapor pressure to form volatile compounds which are then analyzed by gas chromatography. Pyrolytic techniques are a very useful tool for elucidating the structure of macromolecules.

In operation, pyrolysis is quite simple. A sample is placed on a platinum filament. The filament is installed at the inlet of a gas chromatograph and then heated by conduction of an electric current. The volatile compounds formed in this inert atmosphere are passed into the column for analysis.

For pyrolysis to be a useful analytical tool, the pyrolysis and the resulting chromatogram must be reproducible. The most important factors which must be controlled are pyrolysis time and temperature. Other factors which influence reproducibility are type and size of sample, purity of carrier gas and condition of the filament.

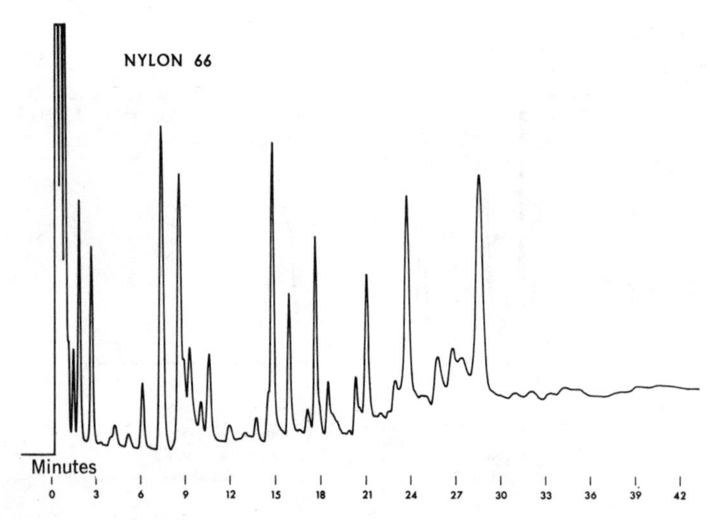

FIGURE XI-19—A TYPICAL PYROLYSIS OF NYLON 66

BIBLIOGRAPHY

1. *Baumann, F., Capillary Columns, Varian Aerograph Reprint W-123.*
2. *Ettre, L.S., Open Tubular Columns, Plenum Press, New York, 1965.*
3. *Schwartz, R.D. and Brasseaux, D.J., Analytical Chemistry, 35, 1374 (1963).*
4. *Scott, C.G., J. Inst. Petroleum 45, 118 (1959).*
5. *Booker, J., Varian Aerograph, personal communication.*
6. *Hollis, O.L. Analytical Chemistry, 38, 309 (1966).*
7. *Klein, Allan, Porapak Columns, Technical Bulletin, 128-66, Varian Aerograph.*

XII . LABORATORY EXERCISES

LABORATORY EXERCISE #1

Determination of Theoretical Plate Height (HETP)

Purpose

To familiarize one with the principle of operation of a gas chromatograph and the effect of the carrier gas flow rate on the theoretical plate height.

Column

The column used in this experiment is a 6 ft. x 1/4 in. column containing 20% X F-1150 (nitrile-silicone oil) on 60/80 mesh Chromosorb W.

Instrument Conditions

Injector Temperature	200°C
Oven Temperature	60°C
Detector Temperature	200°C
Filament Current	170 ma.
Attenuator	32

Procedure

1. Review Chapter III.

2. With the sample provided (Ethylbenzene), inject 1.0 microliter at 5 different flow rates: 25, 50, 75, 100 and 125 ml/min.

3. Calculate theoretical plates by means of the following equation:

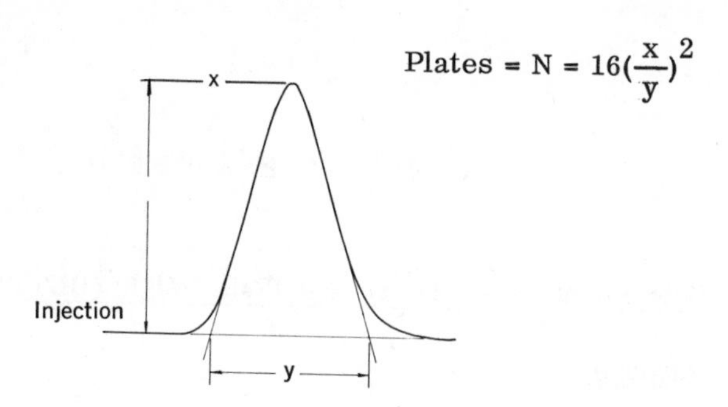

FIGURE 1 —CALCULATION OF THEORETICAL PLATES

4. Calculate the HETP (Height Equivalent to a Theoretical Plate) by the following formula:

$$HETP = \frac{L}{N}$$

Where L is the length of the column expressed in CM

5. Plot the five HETP values obtained vs the flow rates in ml/min. Determine the optimum flow rate for this analysis.

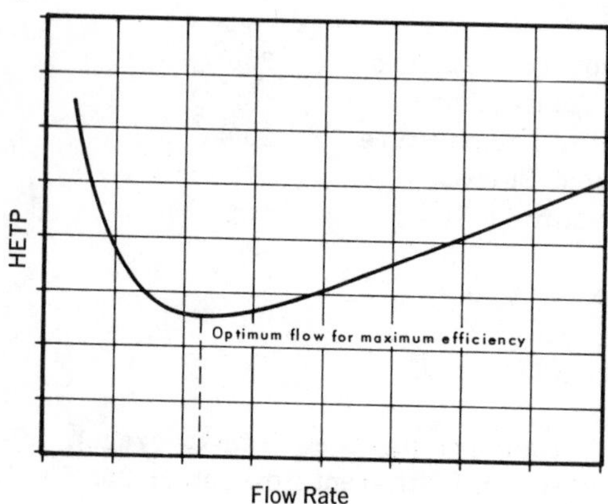

FIGURE -2 —VanDEEMTER PLOT

Discussion

A curve similar to the one shown above should be obtained. In selecting the optimum flow rate, it is preferable to select a flow slightly on the high side. This allows a shorter analysis time without much loss of efficiency. In a difficult separation it may be necessary to select the flow giving maximum efficiency and sacrifice speed.

LABORATORY EXERCISE #2

Qualitative and Quantitative Analysis

Purpose

To identify and determine the concentration of an unknown mixture.

Column

The column used in this experiment is a 6 ft. x 1/4 in. column (for thermal conductivity instruments) or a 6 ft. x 1/8 in. (for flame ionization instruments) containing 20% XF-1150 (nitrile-silicone oil) on 60/80 mesh Chromosorb W.

Instrument Conditions

Injector Temperature	200°C
Oven Temperature	60°C
Detector Temperature	200°C
Filament Current	170 ma. (for TCD)
Range	100 (for FID)
Hydrogen	25 ml/min
Air	300 ml/min

Procedure

1. Review Chapters VI and VII.

2. Adjust the carrier flow rate to optimum as determined by laboratory exercise #1. If exercise #1 has not been performed the flow rate will be preset by instructor.

3. Inject approximately one microliter of the unknown mixture provided. Attenuate as necessary to keep peaks on scale.

4. Inject small amounts (0.5 ul) of the standards provided. Establish the identity of the unknown by comparing retention times.

5. Determine the areas of the individual compounds by the method described by your instructor.

6. Determine the composition calculated from peak areas. This is done by:

$$\%A = \frac{\text{Area A x 100}}{\text{Total Area}}$$

Component	Retention Time (Min)	Area	Area %

Discussion

Retention time can be used to identify unknown components. Peak area can be related to sample concentration by normalization and the use of correction factors.

LABORATORY EXERCISE #3

Programmed Temperature Analysis

Purpose

To illustrate isothermal vs programmed temperature operation as applied to the analysis of a hydrocarbon mixture.

Column

A 5 ft. x 1/8 in. or a 5 ft. x 3/16 in. column containing 5% SE-30 on 60/80 mesh Chromosorb W is used in this experiment.

Instrument Conditions

Injector Temperature	250°C
Oven Temperature	
Isothermal	100°C
Programmed - Initial	60°C
Final	175°C
Rate	10°C/min
Detector Temperature	250°C
Carrier Flow Rate	40 cc/min (for TCD)
	25 cc/min (for FID)
Range	100 (for FID)
Hydrogen	25 cc/min
Air	300 cc/min

Procedure

1. Review Chapter IX.

2. Adjust the oven temperature to 100°C. Allow the column and detector to stabilize.

3. Inject 1 ul of the sample provided. Attenuate as required to keep the peaks on scale.

4. Six peaks should be observed: heptane, octane, decane, undecane, dodecane and tetradecane.

5. Lower the oven temperature to 60°C and allow collumn to stabilize at this temperature.

6. Set program rate at 10°C/min and upper limit at 175° C.

7. Inject 1 ul of sample and immediately start the program. Attenuate as required.

8. When the last component has been eluted, cool the oven and reset at the initial temperature.

Discussion

Compare the isothermal and the programmed temperature runs. Programming the temperature of a G.C. column affords better resolution of complex and/or wide boiling range mixtures. The peaks are more symmetrical and the analysis time is shorter.

LABORATORY EXERCISE #4

Area Measurement

Purpose

To familiarize one with several methods of area measurement.

Procedure

1. Review Chapter VII.

2. Read the disc integrator traces for the H_2S, CH_4, and CO peaks on the chromatogram on page 212. Refer to instructions on reading the integrator trace in the Appendix.

3. Determine the area (in mm^2) of the CO peak by triangulation and height x 1/2 width. Compare the results.

4. The peaks shown on page 213 are triplicate runs of a benzene-toluene mixture. Measure the areas by triangulation, height x 1/2 width and by the Disc Integrator. Then calculate the composition (weight %) and compare your results. Assume the correction factor for both compounds is 1.0, i.e. Area % = Weight %.

$$\% = \frac{\text{Area of component}}{\text{Total area}} \times 100$$

Discussion

Several methods of determining peak areas can be used: Cut and weigh, planimeter, triangulation, height x 1/2 width, or an integrator, either mechanical (Disc) or electronic. The benzene-toluene peaks shown on page 213 were measured by several of these methods. The table shows the results and precision obtained. Compare with your results.

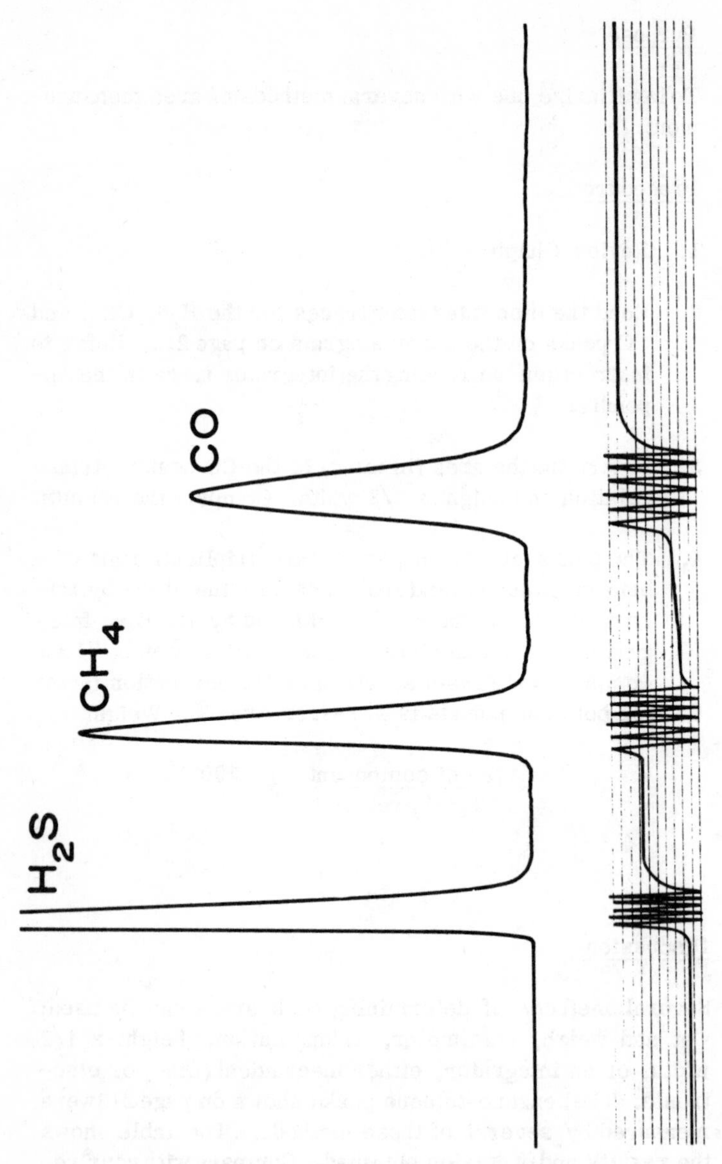

FIGURE III—ANALYSIS OF H_2S, CH_4, CO

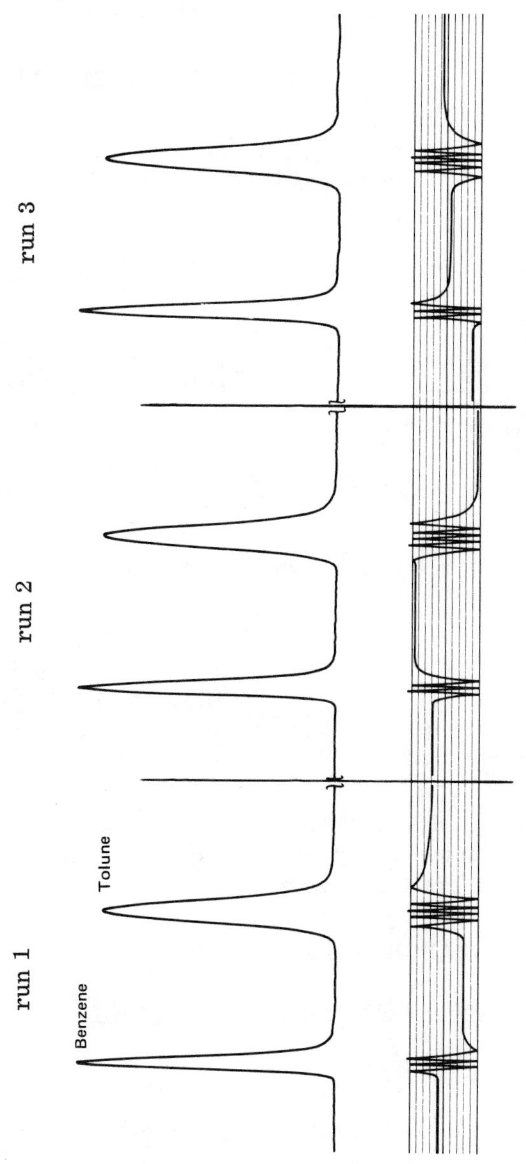

FIGURE IV—TRIPLICATE ANALYSIS OF BENZENE-TOLUENE MIXTURE

TABLE I—COMPARISON OF INTEGRATION METHODS

	#1	#2	#3	Ave.	σ abs	σ ret
Model 475 Digital Integrator						
Benzene	38.425	38.460	38.430	38.438	0.0137	0.0358%
Toluene	61.574	61.539	61.569	61.561	0.0184	0.0299%
Disc Integrator						
Benzene	38.4	38.4	38.7	38.5	0.17	0.45 %
Toluene	61.6	61.6	61.3	61.5	0.17	0.28 %
Triangulation						
Benzene	38.7	37.9	37.2	37.9	0.75	1.97 %
Toluene	61.3	62.1	62.8	62.1	0.76	1.22 %
Height x Width at 1/2 Height						
Benzene	37.5	38.8	37.2	37.8	0.85	2.25 %
Toluene	62.5	61.2	61.8	61.8	0.92	1.49 %
Weighing Paper						
Benzene	40.8	39.5	38.9	39.7	0.97	2.45 %
Toluene	59.2	60.5	61.1	60.3	0.97	1.54 %
Planimeter						
Benzene	38.2	37.3	39.2	38.2	0.95	2.49 %
Toluene	61.3	62.7	60.6	61.5	1.07	1.75 %

Electronic digital and Disc integration are the most precise methods of measuring peak areas; weighing paper and the planimeter are less precise.

LABORATORY EXERCISE #5

Retention Times for Members of a Homologous Series

Purpose

To illustrate the use of retention time plots for qualitative analysis.

Column

The column used in this experiment is a 6 ft. x 1/4 in. or 6 ft. x 1/8 in. column containing 20% XF-1150 (nitrile-silicone) on 60/80 mesh Chromosorb W.

Instrument Conditions

Injector Temperature	$200^{o}C$
Oven Temperature	$60^{o}C$
Detector Temperature	$200^{o}C$
Carrier Flow Rate	60 cc/min for 1/4 in.column
	25 cc/min for 1/8 in.column
Filament Current	170 ma. (for TCD)
Range	100 (for FID)
Hydrogen	25 ml/min
Air	300 ml/min

Procedure

1. Review the Chapter on Qualitative Analysis.

2. Inject 1 ul of the sample provided. Mark the point of injection. If possible, use a fast chart speed to allow accurate measurement of retention times. The compounds will be eluted in order of increasing number of carbon atoms.

3. Measure the adjusted retention time from the air peak for each component in the mixture.

4. Plot the log of retention time versus number of carbon atoms. Draw the best straight line through the points.

5. Inject the unknown and identify it by using the retention time plot.

Discussion

The logarithmic relationship between retention volume and retention time for a constant flow rate) and the number of carbon atoms in the compounds of a homologous series was shown in the first publication on gas chromatography and in many since then. In some cases the relationship also holds between retention volume and boiling point or molecular weight.

Under a given set of conditions, (temperature, flow and column), a plot can be made for each type of compound which can be separated on the column being used. Once the plots are made for the lower members of several series, the straight lines can be extrapolated to permit a fast estimate to be made of the identity of an unknown.

LABORATORY EXERCISE #6

Injection Technique and Introduction to Statistics

Purpose

To introduce beginners to the technique of syringe injection, and to illustrate the precision (repeatability) possible.

Column

The column used in this experiment is a 6 ft. x 1/4 in. or 6 ft. x 1/8 in. column containing 20% XF-1150 (nitrile-silicone) on 60/80 mesh Chromosorb W.

Instrument Conditions

Injector Temperature	$200^{\circ}C$
Oven Temperature	$60^{\circ}C$
Detector Temperature	$200^{\circ}C$

Carrier Flow Rate	As determined in Ex. #1 or 25 ml/min for 1/8" column
Filament Current	170 ma (for TCD)
Range	100 (for FID)
Hydrogen	25 ml/min
Air	300 ml/min

Procedure

1. Check column temperature. Adjust flow rate to optimum flow determined in Exercise #1. If exercise #1 has not been done, instructor will have preset carrier flow.

2. Inject 1 ul of sample provided using a 10 ul syringe. Attenuate so that the peak height is between 50 and 90% of the recorder scale.

3. Inject three replicate samples, one after the other, at one minute intervals.

4. Measure the peak height in millimeters and calculate: Average height; individual error; relative average error.

Example

Peak heights - 140 mm
142
144
426

Average peak height=426/3 = 142

Individual error = 142 - 140 = 2 mm
142 - 142 = 0
142 - 144 = 2

Average error 4/3 = 1.3 mm

$$\text{Relative average error} = \frac{1.3}{142} \times 100 = 0.9\%$$

5. If relative average error exceeds 3%, repeat the experiment.

Discussion

In gas chromatography several factors contribute to errors in quantitative analyses. One of these is operator skill in the introduction of samples. To be able to obtain the best precision for quantitative analysis, one must have a proper injection technique.

LABORATORY EXERCISE #7

Determination of Ethanol in Water by Internal Standardization

Purpose

To illustrate the use of a hydrogen flame detector in the analysis of water samples.

Column

A 15 ft. x 1/8 in. column containing 10% Castorwax on Chromosorb T.

Instrument Conditions

Injector Temperature	170°C
Oven Temperature	100°C
Detector Temperature	185°C
Carrier Flow Rate	22 ml/min
Hydrogen Flow Rate	30 ml/min
Air Flow Rate	300
Range	1
Attenuation	2

Procedure

1. Pipet 1 ml of the methyl ethyl ketone standard solution into a small bottle. Stopper until a 1 ml sample of an ethanol-water standard has been added. Mix thoroughly.

2. Inject 1 ul of this solution into the chromatograph. Run a duplicate sample.

3. Repeat steps 1 and 2 with each of the standard solutions and with the unknown.

4. Measure the peak heights in millimeters or use an integrator to obtain peak areas.

5. Plot $\frac{\text{peak height ethanol}}{\text{peak height MEK}}$ vs $\frac{\text{weight \% ethanol}}{\text{weight \% MEK}}$

   or

   $\frac{\text{peak area ethanol}}{\text{peak area MEK}}$ vs $\frac{\text{weight \% ethanol}}{\text{weight \% MEK}}$

6. Draw the best straight line through the points plotted. Read weight ratio for the unknown. Calculate % ethanol in the unknown.

Discussion

The hydrogen flame is insensitive to water; thus it is the most convenient detector to use when determining low concentrations of organics in water.

APPENDIX

A. USE OF SYRINGES IN INJECTING SAMPLES

In filling a microliter syringe with liquid it is usually desirable to exclude all air initially. This can be accomplished by repeatedly drawing liquid into the syringe and rapidly expelling it into the liquid. Viscous liquids must be drawn into the syringe slowly; very fast expulsion of a viscous liquid could split the syringe. Draw up about twice as much liquid into the syringe as you plan to inject.

The following procedure should be used to adjust the volume of a liquid in the syringe. Hold the syringe vertically with the needle pointing up. Put the needle through a tissue so that liquid which is expelled will be absorbed in it. Any air still in the syringe should go to the top of the barrel. Push the plunger until it reads the desired value. The excess air should have been expelled. Wipe off the needle with the tissue on it. Draw some air into the syringe now that the exact volume of liquid has been measured. This will serve two purposes: first, the air will give a peak on the chromatogram which will be used to calculate "adjusted" retention volumes; second, the air prevents any liquid from being expelled if the plunger is accidentally pushed.

1. Injection procedure

Hold the syringe in two hands. Use one (normally the left to guide the needle into the septum, and the other to provide force to pierce the septum and to prevent the plunger from being forced out by the pressure in the gas chromatograph (use the thumb on your right hand). This latter point is very important when large volumes are to be injected (i.e., gas samples) or when the inlet pressure is extremely high. Under these conditions, if care is not exercised the plunger will be blown out of the syringe.

Insert the needle through the septum and as far into the injection port as possible, depress the plunger, hesitate a second or two, then withdraw the needle (keeping the plunger depressed) as rapidly and smoothly as possible.

2. Cleaning Procedure

When high-boiling liquids are being used, the syringe should be washed out with a volatile solvent like methylene chloride, acetone, etc. This can be done by repeatedly pulling the wash liquid into the syringe. Then remove the plunger and dry the syringe by pulling air through it. Attach the barrel of the syringe to a vacuum pump (with appropriate trap), or a water aspirator. This way air is pulled in through the needle and dust cannot get into the barrel to clog it. Wipe the plunger with a tissue and reinsert. If the needle gets dulled after long use, it can be sharpened on a small grindstone. Observe all of the directions on the sheet supplied with the syringe by the manufacturer.

B. MEASUREMENT OF FLOW RATE

The soap bubble flow meter should be attached to the outlet from the gas chromatograph.

Put soap solution in the flow meter, filling the rubber bulb and the tube up to, but not covering, the side arm. Add enough solution so that it covers the side arm only when the rubber bulb is squeezed. Then, when carrier gas is flowing and the bulb is squeezed, a soap bubble will be formed and carried up the graduated tube. Wet the tube thoroughly with several bubbles before measuring.

To measure the flow rate, use a stop watch to determine how long it takes the soap bubble to travel the 10 ml between the markings on the calibrated tube. Calculate the flow rate in millimeters per minute by dividing 600 by the time in seconds.

$$\text{Flow rate (ml/min)} = \frac{600}{\text{seconds}}$$

C. COLUMN PREPARATION

Several methods of coating the solid support with liquid can be used.

1. Rotating evaporator method

   The correct amount of liquid dissolved in a suitable solvent is placed in a round bottom flask. The weighed amount of solid support is added.

   The flask is connected to the rotating evaporator. A water aspirator is used to reduce pressure in the flask. The flask is rotated until all of the solvent is evaporated. Use of a heat lamp helps evaporation. This method is not recommended for use with Chromosorb T.

2. Pan coating method

   The weighed amount of liquid phase dissolved in the correct amount of solvent is added to the weighed solid support in a pan. The amount of solvent used is just enough to wet the solid support with no excess. The solvent is allowed to evaporate spontaneously or with judicious application of heat. The mixture is gently agitated during drying by shaking the pan. Do not stir, as this may crush the Chromosorb particles. The amount of solvent to use may be calculated from the following table. Milliters of solvent required is obtained by multiplying the grams of support by the **Z** factor.

TABLE I—AMOUNT OF SOLVENT TO USE WITH SUPPORTS

Support	ml solvent/ g support Z	Example for 20gram use
Chromosorb P	1.5	30 ml
Chromosorb W	2.0	40 ml
Chromosorb G	0.5	10 ml
Chromosorb T	1.0	20 ml
Firebrick	1.5	30 ml
Fluoropak 80	0.8	16 ml

3. Funnel coating method

Add 20 gm of support to 100 ml of the proper solution of the liquid phase in a filter flask. The pressure in the flask is reduced for a few minutes with a water aspirator. Then release the pressure and let the flask stand for 15 minutes. Pour the material into a sintered glass funnel, and let it drain freely until the support settles. Apply vacuum for approximately 5 minutes. Spread the support on filter paper to dry. After air drying, dry the coated support in an oven at 80-100° C. Low temperature liquid phases should be dried at room temperature. Do not resieve before use.

The correct solution to obtain a given weight coating on the support may be calculated from the following table. % solution = % coating multiplied by ϕ factor.

TABLE II—FACTORS FOR CALCULATING SOLUTION PERCENTAGES

Support	ϕ Factor per 1%	Example (for 10% liquid phase) Solution by weight
Chromosorb P	0.75	7.5%
Chromosorb W	0.55	5.5
Chromosorb G	1.15	11.5
Chromosorb T	1.50	15.0
Firebrick	0.75	7.5
Fluoropak 80	2.00	20.0

4. Packing the column
   A straight piece of tubing of the desired length and diameter is plugged loosely at one end with a small piece of glass wool. A funnel is attached to the open end and filled with coated support. The column is vibrated with a hand vibrator or tapped with a stick until no further packing can be added. The column is bounced lightly on the floor several times to aid settling of the support. Then the open end of the column is plugged. The column may then be coiled or bent to fit the configuration of the chromatograph oven.

5. Conditioning the column
   Columns should be conditioned for at least 2 hours at 25° C above the maximum temperature at which the column will be used, but below the maximum temperature limit for the liquid phase. A small carrier gas flow (5-10 ml/min) should be used during conditioning. The exit end of the column should be left disconnected from the detector, to avoid contaminating the detector.

   Special conditioning instructions for specific liquid phases follow:

STAP--Condition at 250° C overnight with no flow. Then lower the temperature to 225° C, turn on the carrier gas flow and allow to remain at these conditions until stable.

FFAP--Condition at 100° C for two days with no flow. Then condition at operating temperature with flow until stable.

SE-30 --Condition at 250° C overnight with no flow. Then condition at operating temperature with flow until stable.

D. MAXIMUM TEMPERATURES AND SOLVENTS FOR LIQUID PHASES

These are the maximum allowable temperatures for columns used with a thermal conductivity detector; a higher temperature will destroy the columns. However, the maximum usable temperatures for columns with ionization detectors may be as much as 100° C lower.

Solvent Code

1 - Acetone	4 - Methylene Chloride	8 - Carbon Disulfide
2 - Benzene	5 - Ethyl Acetate	9 - Water
3 - Chloroform	6 - Methanol	h - Hot
	7 - Toluene	

Liquid Phase	Max. Temp. °C	Sol- Vent	Liquid Phase	Max. Temp. °C	Sol- Vent
Acetonyl Acetone			Alkaterge T	75	3&4h
(2,5 Hexanedione)	25	1	Amine 220	180	3&4
Adiponitrile	50	3&4	Apiezon H	275	3

Liquid Phase	Max Temp °C	Sol-Vent
Apiezon J	300	3&4
Apiezon L	300	3&4
Apiezon M	275	3&4
Apiezon N	300	3&4
Armeen SD	100	3&4
Aroclor 1254 (Chlorinated Biphenyl)	125	3&4
Asphalt	300	3&4
Bentone 34	200	4
7,8 Benzoquinoline	150	3&4
Benzyl Cellosolve	50	3&4
Benzyl Cyanide (Phenyl Acetonitrile)	35	3&4
Benzyl Cyanide-Silver Nitrate	25	4
Benzyldiphenyl	140	4
Benzyl Ether	50	3&4
Bis (2 Ethylhexyl Tetrachlorophathalate)	150	3&4
Bis 2-Methoxy Ethyl Adipate	150	4
Bis (2(2-Methoxyethoxy)Ethyl) Ether	50	3&4
Butanediol Adipate	225	3&4
Butanediol Succinate	225	3&4
Carbowax 20M	250	3&4
Carbowax 20M TPA	250	4&6
Carbowax 300	100	3&4
Carbowax 400	125	3&4
Carbowax 400 Monooleate	125	3&4
Carbowax 550	125	3&4
Carbowax 600	125	3&4
Carbowax 600 Monostearate	125	3h&4h
Carbowax 750	150	3&4
Carbowax 1000	175	3&4
Carbowax 1500	200	3&4
Carbowax 1540	200	3&4
Carbowax 4000	200	3&4
Carbowax 4000 Monostearate	220	4
Carbowax 6000	200	3&4
Castorwax	200	3&4
Celanese Ester #9	200	3&4
Chloronaphthalene	75	3&4
Cyanoethylsucrose	175	4
n-Decane	30	1&4
Dibutyl Maleate	50	3&4
Dibutyl Phthalate	100	3&4
Dibutyl Tetrachlorophthalate	150	4
Didecyl Phthalate	125	3&4
Diethylene Glycol Adipate (DEGA)	190	3&4
Diethylene Glycol Sebacate (DEGSE)	190	3h&4
Diethylene Glycol Succinate (DEGS)	190	1&4
Di(2-Ethylhexyl) Sebacate	125	3&4
Diethyl Sebacate-Sebacic Acid	75	4
d-Diethyl tartrate	125	3&4
Diglycerol	120	6&9
Diisodecyl Phthalate	175	3&4
Diisoactyl Sebacate	175	3&4
Dimer Acid	150	3&4
2,4 Dimethylsulfolane	50	3&4
Dimethylsulfoxide		
Dinonyl Phthalate	175	3&4

Liquid Phase	Max Temp °C	Sol- Vent
Dioctyl Phthalate	175	3&4
Dioctyl Sebacate	100	3&4
Dowfax 9N9	225	6
Ditetrahydrofur- furyl Phthalate	125	3&4
EPON Resin 1001	225	3h&4h
Ethofat 60/25	140	3h&4h
Ethomeen	75	3&4
Ethylene Glycol Adipate (EGA)	200	3h&4h
Ethylene Glycol Isophthalate (EGIP)	250	3h&4h
Ethylene Glycol Sebacate (EGSE)	200	3h&4h
Ethylene Glycol Succinate (EGS)	200	3h&4h
FFAP	275	4
Flexol Plasticizer 8N8	180	3&4
Fluorolube GR 362	100	2h
Glycerol	100	6&9
Hallcomid M-18	150	3&4
Hallcomid M-18 OL	150	3&4
n-Hexadecane	50	3&4
n-Hexadecene	50	3&4
HHK Mix	50	3&4
Hexamethylphos- phoramide (HMPA)	50	3&4
2, 5 Hexanedione (see Acetonyl Acetone)		
Hyprose SP-80	190	3h&4h
IGEPAL (Nonyl Phenoxypolyox- yethylene Ethanol)	200	3h&4h
Ionox 330	250	4
Isoquinoline	50	3&4
Kel F Grease	200	3&4
Kel F Oil #3	100	3&4
Kel F Oil #10	100	3&4

Liquid Phase	Max Temp °C	Sol- Vent
LAC 1-R-296 (see DEGA- Diethylene Glycol Adipate)		
LAC 2- -446 (DEGA cross-linked)	190	1
LAC 3-R-728 (see DEGS- Diethy- lene Glycol Succinate)		
LAC 4-R-886 (see EGS- Ethylene Glycol Succinate)		
Lexan (Polycarbonate resin)	300	hot DMF 3h
Lithium Chloride	500	9
Mannitol	200	6&9
Neopentyl Glycol Adipate (NPGA	240	3&4
Neopentyl Glycol Isophthalate (NPGIP)	240	3
Neopentyl Glycol Sebacate (NPGSE)	240	3&4
Neopentyl Glycol Succinate (NPGS)	240	3&4
Nitropimelon- itrile (gamma-methyl- gamma)	110	4
Nonyl Phenol	125	4h
Nonyl Phenoxy- polyoxyethylene Ethanol (see IGEPAL)		
Nujol (Mineral Oil)	200	3&4

Liquid Phase	Max Temp ^{o}C	Solvent
b, b' Oxidipro-prionitrile	100	4
Phenyl Acetonitrile (see Benzyl Cyanide)		
Phenyl Diethanol-amine Succinate (PDEAS)	225	1&4
Polyethylene Glycol (see Carbowaxes)		
Poly m-Phenyl Ether (5-ring)	250	3&4
Polypropylene Glycol	150	3&4
Polypropylene Glycol Silver Nitrate	75	3&4
Porapak P, Q, R, S, N	240	
Porapak T	200	
Propylene Car-bonate (1, 2 Propanediol Cyclic Carbonate)	60	4
Quadrol	150	3&4
Reoplex 400 (poly-propylene Glycol Adipate)	190	3&4
Silicone Dow Corning 11	300	2&5
Silicone Dow Corning 200	250	3&4
Silicone Dow Corning 550	275	3&4
Silicone Dow Corning 560 (F60)	250	3&4
Silicone Dow Corning 703	225	3&4
Silicone Dow Corning 710	300	3&4

Liquid Phase	Max Temp. ^{o}C	Solvent
Silicone (Fluoro) QF-1 (FS 1265)	250	1
Silicone Fluid XF-1112 (Nitrile)	200	3&4
Silicone Fluid XF-1125 (Nitrile)	200	3&4
Silicone Fluid XF-1150 (Nitrile)	200	3&4
Silicone GE SF-96	300	3&4
Silicone GE Xe-60 (Nitrile Gum)	275	3&4
Silicone GE Versilube F-50	300	3&4
Silicone Gum Rubber SE-30 (Methyl)	375	3h&4h
Silicone Gum Rubber SE-52 (Phenyl)	300	3h&4h
Silicone Gum Rubber SE-54 (Methyl Phenyl Vinyl)	300	4
Silicone Rubber	250	4
Span 80 (Sorbitan Mono-oleate)	150	3&4
Squalane	30	3&4
Squalene	30	3&4
STAP (Steroid Analysis Phase)	250	1
Sucrose Acetate Isobutyrate (SAIB)	225	3h&4
Tergitol Nonionic NP-35	200	3&4
Tetracyanoethylated Pentaerythritol	180	4
Tetraethylene Glycol	70	4

Liquid Phase	Max. Temp. °C	Sol-Vent
Tetraethylene Glycol Dimethyl Ether (see Bis (2(Methoxyethoxy) Ethyl) Ether)		
Tetraethylene-pentamine	150	4
Tetrahydroxy-ethylethylene-diamine (THEED)	150	3&4
β,β'Thiodipropionitrile	100	3&4
Tributyl Phosphate	40	
Tricresyl Phosphate (TCP) (Tritolyl Phosphate)	125	3&4
Triethanolamine	75	3&4
Trimer Acid	200	3&4
1,2,3, Tris(2 Cyano Ethoxy) Propane (TCEP)	180	3&4

Liquid Phase	Max Temp °C	Sol-Vent
Tritolyl Phosphate (see Tricresyl Phosphate)		
Triton X-305	200	1
TWEEN 80 (Polyoxyethylene Sorbitan Monooleate)	150	3&4
UCON 50 HB 280X Polar	200	3&4
UCON 50 HB 2000 Polar	200	3&4
UCON 50 HB	200	3&4
UCON 50 LB 550X	200	3&4
UCON LB 1715 Non-polar	200	3&4
Versamid 900	250	*
Xylenyl Phosphate-2,4	174	4
Zonyl E-7	200	4

**1:1 butanol & phenol or 87% Chloroform & 13% Methanol. Versamid is unstable at high temperatures in the presence of oxygen.*

E. CHEMICAL STRUCTURES OF COMMON LIQUID PHASES

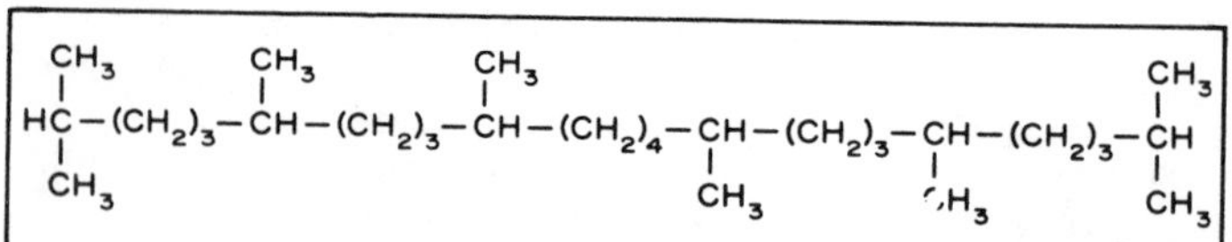

$$(CH_3)_2CH-(CH_2)_3-CH(CH_3)-(CH_2)_3-CH(CH_3)-(CH_2)_4-CH(CH_3)-(CH_2)_3-CH(CH_3)-(CH_2)_3-CH(CH_3)_2$$

SQUALANE

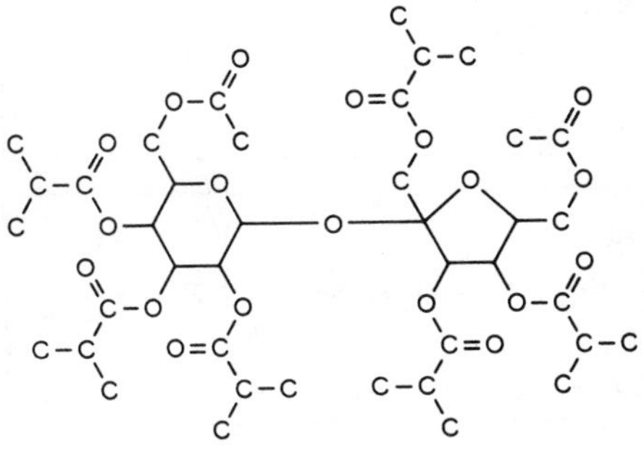

SAIB

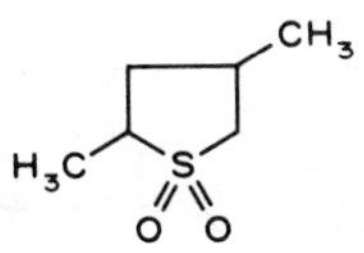

DIMETHYLSULFOLANE

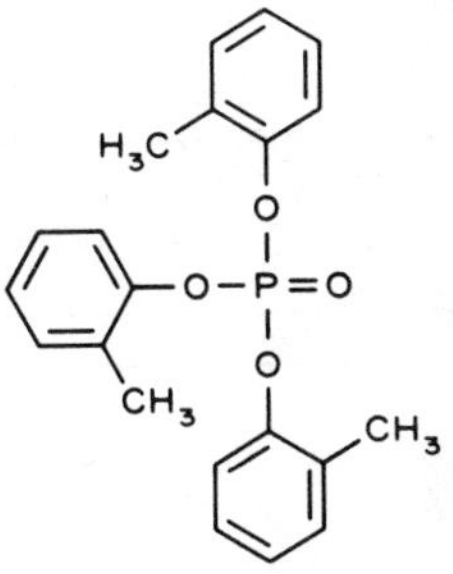

TRICRESYLPHOSPHATE

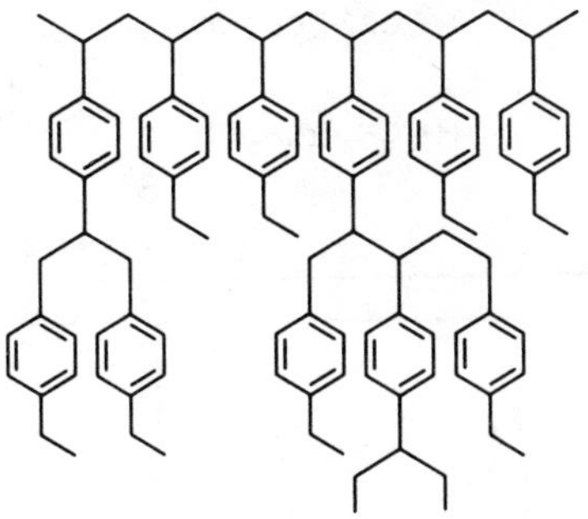

PORAPAK

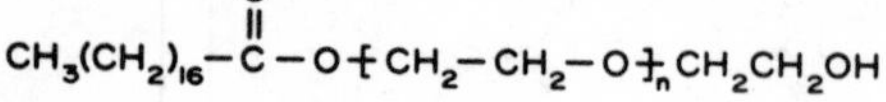

$$CH_3(CH_2)_{16}-\overset{O}{\overset{\|}{C}}-O\left[CH_2-CH_2-O\right]_nCH_2CH_2OH$$

ETHOFAT

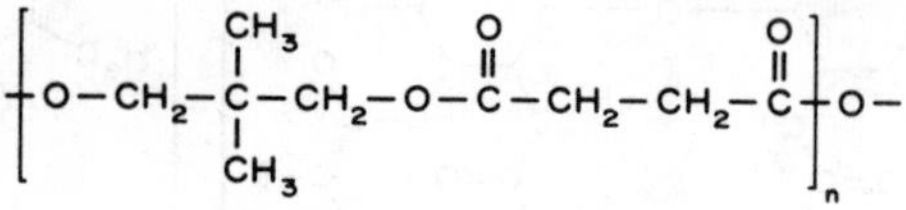

$$\left[O-CH_2-\underset{CH_3}{\overset{CH_3}{C}}-CH_2-O-\overset{O}{\overset{\|}{C}}-CH_2-CH_2-\overset{O}{\overset{\|}{C}}\right]_n O-$$

NEOPENTYL GLYCOL SUCCINATE

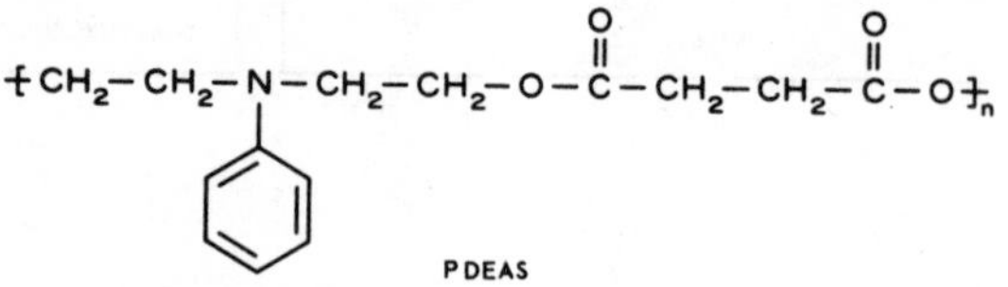

$$\left[CH_2-CH_2-\underset{C_6H_5}{N}-CH_2-CH_2-O-\overset{O}{\overset{\|}{C}}-CH_2-CH_2-\overset{O}{\overset{\|}{C}}-O\right]_n$$

PDEAS

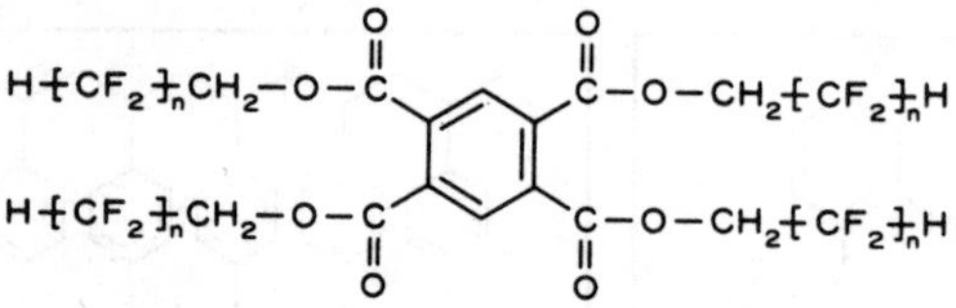

$$H\left[CF_2\right]_nCH_2-O-\overset{O}{\overset{\|}{C}}-C_6H_2(-\overset{O}{\overset{\|}{C}}-O-CH_2\left[CF_2\right]_nH)_2(-\overset{O}{\overset{\|}{C}}-O-CH_2\left[CF_2\right]_nH)$$

$$H\left[CF_2\right]_nCH_2-O-C(=O)- \qquad -C(=O)-O-CH_2\left[CF_2\right]_nH$$

ZONYL E-7

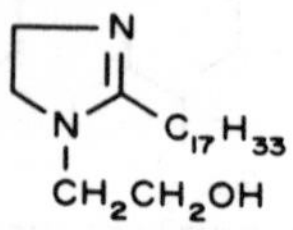

$$N, C_{17}H_{33}, CH_2CH_2OH$$

AMINE 220

$$Cl, Cl, Cl, Cl, \quad \overset{O}{\overset{\|}{C}}-O-C_4H_9, \quad C(=O)-O-C_4H_9$$

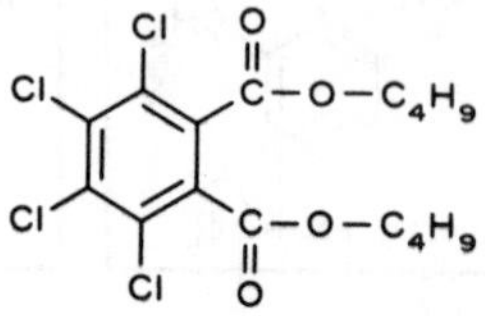

DIBUTYLTETRACHLOROPHTHALATE

$CF_3 \leftarrow CF_2 \rightarrow_n CF_3$

KEL F OIL No. 10, No. 3

$Cl \leftarrow CF_2CFCl \rightarrow_n Cl$

KEL F GREASE

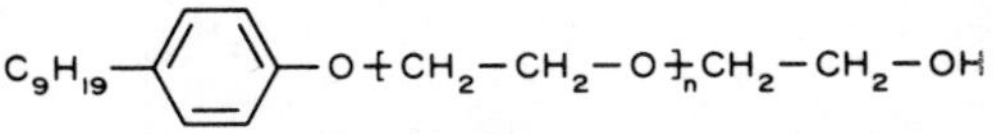

$C_9H_{19}-C_6H_4-O \leftarrow CH_2-CH_2-O \rightarrow_n CH_2-CH_2-OH$

IGEPAL

$\leftarrow CH_2-CH_2-O-CH_2-CH_2-O-C(=O)-CH_2-CH_2-C(=O)-O \rightarrow_n$

DEGS

$CH_3(CH_2)_5-CH(OH)-CH_2-CH=CH-(CH_2)_{17}C(=O)OH$

CASTORWAX

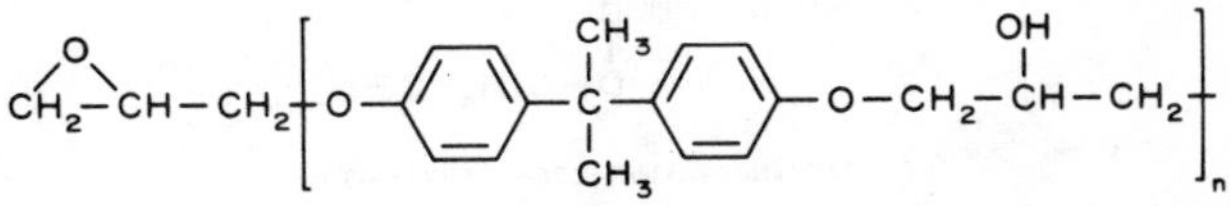

$CH_2(O)CH-CH_2 \left[O-C_6H_4-C(CH_3)_2-C_6H_4-O-CH_2-CH(OH)-CH_2 \right]_n$

EPON 1001

$\left[O-C_6H_4-C(CH_3)_2-C_6H_4-O-C(=O) \right]_n$

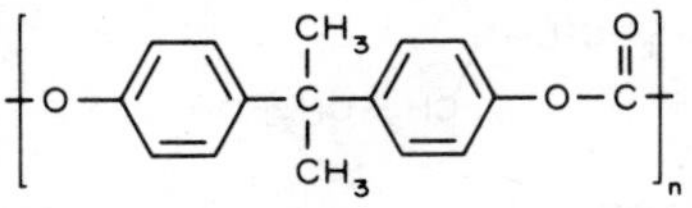

LEXAN

$OH \leftarrow CH_2-CH_2-O \rightarrow_n H$

CARBOWAX

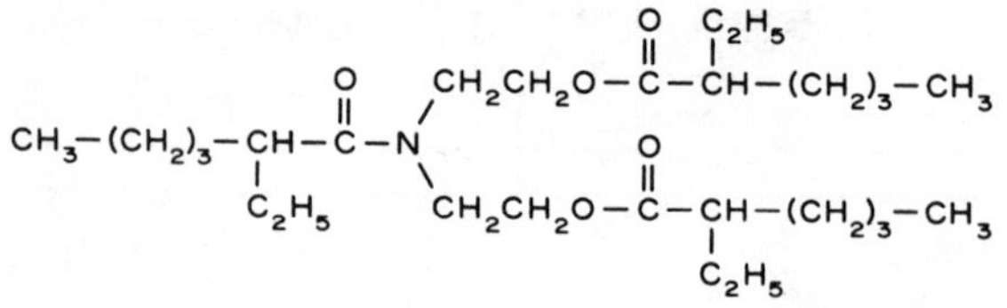

$$CH_3-(CH_2)_3-CH(C_2H_5)-C(=O)-N[CH_2CH_2O-C(=O)-CH(C_2H_5)-(CH_2)_3-CH_3]_2$$

FLEXOL 8N8

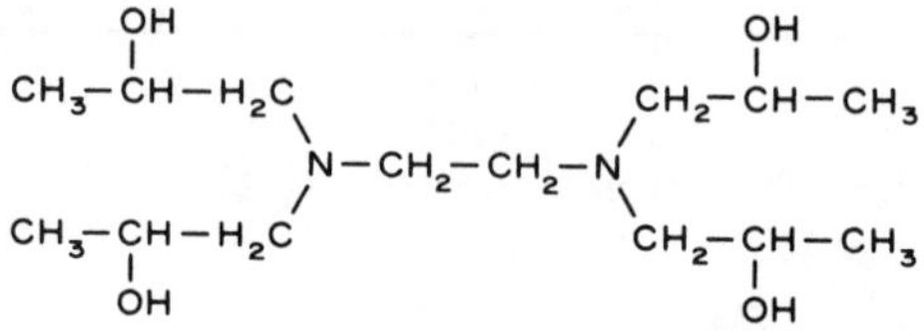

$$[CH_3-CH(OH)-H_2C]_2N-CH_2-CH_2-N[CH_2-CH(OH)-CH_3]_2$$

QUADROL

$$C[C-O-C_2H_4-C\equiv N]_4$$

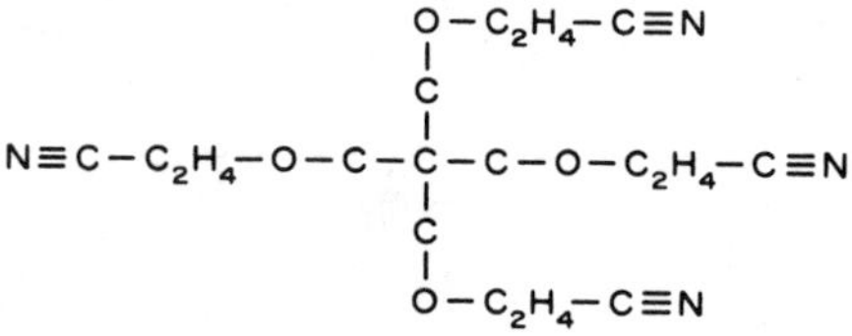

TETRACYANOETHYLATEDPENTAERYTHRITOL

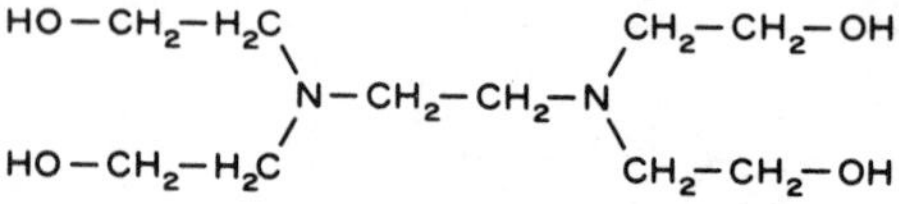

$$[HO-CH_2-H_2C]_2N-CH_2-CH_2-N[CH_2-CH_2-OH]_2$$

THEED

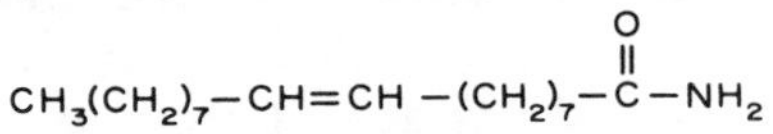

$$CH_3(CH_2)_7-CH=CH-(CH_2)_7-\overset{O}{\overset{\|}{C}}-NH_2$$

HALL COMID M 18OL

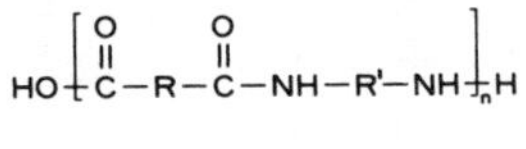

$$HO\left[\overset{O}{\overset{\|}{C}}-R-\overset{O}{\overset{\|}{C}}-NH-R'-NH\right]_n H$$

VERSAMID

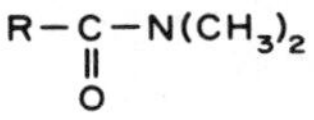

$$R-\underset{O}{\underset{\|}{C}}-N(CH_3)_2$$

HALL COMID M 18

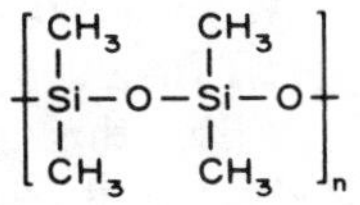

$$\left[\begin{array}{c}CH_3 \quad\quad CH_3\\ |\quad\quad\quad\quad | \\ Si-O-Si-O \\ |\quad\quad\quad\quad | \\ CH_3 \quad\quad CH_3\end{array}\right]_n$$

SE-30 DOW-200
DOW-11 SF-96

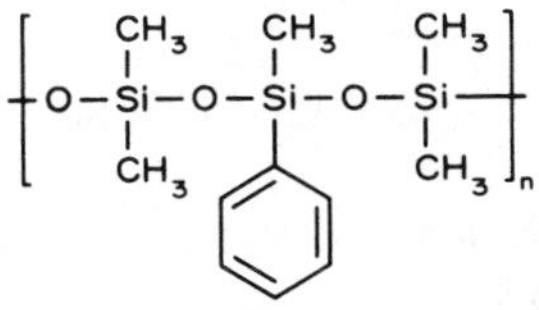

$$\left[-O-\underset{CH_3}{\overset{CH_3}{Si}}-O-\underset{C_6H_5}{\overset{CH_3}{Si}}-O-\underset{CH_3}{\overset{CH_3}{Si}}-\right]_n$$

SE-52
DOW-710
DOW-550

$$Si(CH_3)_3-O\left[\begin{array}{c}CH_3\\ |\\ Si-O\\ |\\ CH_2\\ |\\ CH_2\\ |\\ C\equiv N\end{array}\right]_n Si(CH_3)_3$$

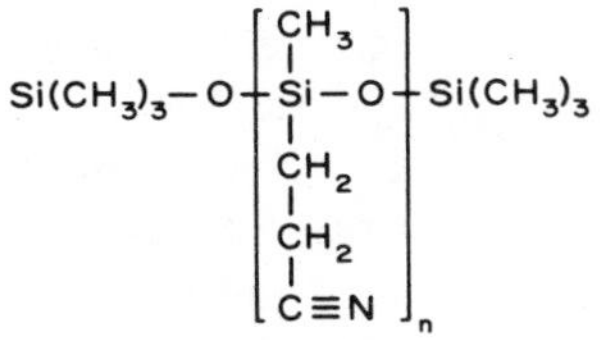

XF-1150 XF-1112
XE-60 XF-1125

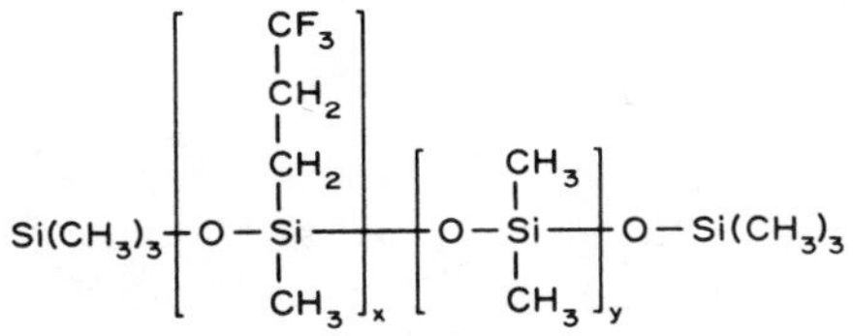

$$Si(CH_3)_3\left[O-\begin{array}{c}CF_3\\ |\\ CH_2\\ |\\ CH_2\\ |\\ Si\\ |\\ CH_3\end{array}\right]_x\left[O-\underset{CH_3}{\overset{CH_3}{Si}}\right]_y O-Si(CH_3)_3$$

QF-1 (FS-1265)

F. PROCEDURE FOR PREPARING AEROPAK 30

This technique for treating Chromosorb W was first described by Horning and co-workers. (Reference 9, Chapter IV).

1. Resieve the support as purchased from supplier. This removes fines produced in shipment.
2. To 30 gm support add excess con. HCl (Baker) and allow it to remain overnight.
3. Decant, repeat 3 or 4 times, 1 hour each.
4. Wash 8-10 times with deionized water. Decant off fines.
5. Filter through sintered glass funnel.
6. Wash with excess methanol.
7. Wash with excess acetone.
8. Oven dry at 80-90°C, after a preliminary air dry (30 minutes).
9. Immediately silanize by adding 30 gm support to 200 cc of 5% Dimethyldichlorosilane in toluene in an Erlenmeyer flask.
10. Draw vacuum to remove air bubbles and trapped air from surface of support.
11. Let stand for 5 minutes.
12. Filter through sintered glass funnel.

CAUTION

Do not allow air to be sucked through support at this point.

13. Wash with 500 ml toluene. Decant off fines.
14. Wash with excess methanol. Decant off fines.

In all washing procedures avoid excess stirring or mechanical agitation which will fracture the Chromosorb.

15. Suck dry. Place on filter paper. Place in 80-90°C oven after a preliminary air dry.

G. HOW TO READ A DISC INTEGRATOR PEN TRACE

1. To read the integrator pen trace, first establish the desired time interval from the recorder pen trace and then project directly down to the integrator pen trace. The value of that interval is then obtained by counting the chart graduations (not time lines) crossed by the integrator pen trace. Every division has an arbitrary value of 10. It will be seen that a complete traverse would have a value of 100. Also note that the space between "blips" is equivalent to 600 counts.

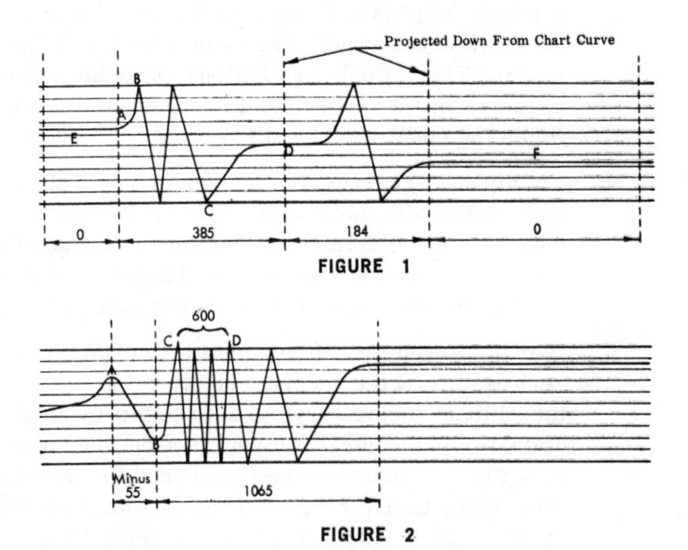

FIGURE 1

FIGURE 2

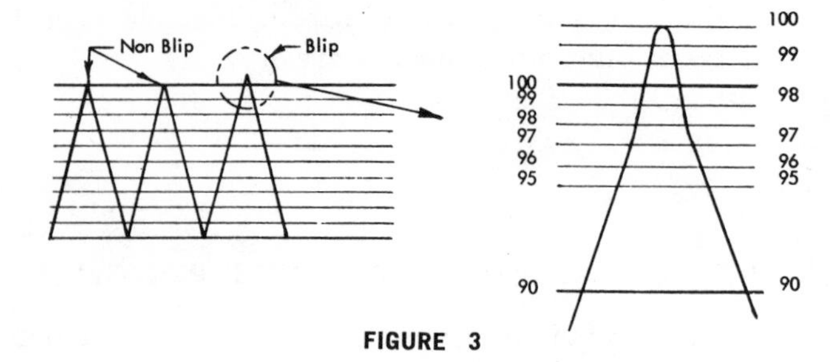

FIGURE 3

2. With care a partial traverse can be read to within one count. For example, the time interval between A and B (Figure 1) can be either read as 35 or 36. Obviously it is not 37 or 34. When the traverse is close to or on a chart graduation, it is possible to read it exact. For example, interval C to D is exactly 50 counts.

3. A small "blip" (see C and D in Figure 2) is drawn every sixth traverse. The value between "blips" is 600. The "blips", in addition to speeding up the reading, make it possible to record up to three times the counts per length of chart.

4. For highest accuracy, the integrator zero* should correspond with the electrical zero. It is not important that the integrator zero correspond with the chart zero since the chart no longer has quantitative significance when the integrator is used.

5. If it is not convenient to maintain the integrator zero exactly at the recorder electrical zero, or if the shift is momentary, it is possible by an examination of the integrator pen trace between chart peaks to estimate an offset in counts per minute. This value times the peak widths in minutes can be added to or subtracted from the peak area.

* *The integrator is at zero when the pen is motionless (draws a straight line, see E and F, (Figure 1). It is rare that the integrator pen will completely stop since the integrator is normally more sensitive at zero than the recorder.*

6. A negative value of integration can be detected by noting when the integrator pen has reversed motion other than at the extremes of travel (see A and B in Figure 2). (There is a possibility that the negative reversal could occur simultaneously with the normal reversal.)

7. Project down from the "null" or minimum point between the peaks for unresolved peaks. The area obtained will closely approximate the actual peak area, particularly if the peaks are similar in shape.

8. Figure 3 illustrates how to read the pen trace in the "blip". Up to about 97 counts, the values are read in the normal manner. From 97 to 100 counts the scale has been doubled as shown. Accordingly, the trace normally read as 100 counts would in the "blip" be read as 98 or 99 counts, etc.

H. RECOMMENDED LITERATURE

1. Textbooks

a. D. AMBROSE and B. A. AMBROSE: Gas Chromatography, G. Newnes Ltd., London - Van Nostrand Co., Princeton, New Jersey, 1961.
(Authors have provided a compact balanced primer for the beginner. They selected essential features of the techniques and discussed them with enough material to be clear. Suggested items for further reading appear at the end of each chapter.)

b. E. BAYER: Gas Chromatography. Elsevier, Amersterdam, New York, 1961.
(Extremely useful laboratory handbook. It has a good balance of history, theoretical background and the practical aspects. Very good coverage, but brief.)

c. S. DAL NOGARE and R. S. JUVET: Gas Chromatography - Theory and Practice, Interscience - J. Wiley & Sons, New York, 1962.
(Primarily for the experienced chromatographer. Theory and practice, -- valuable as a reference book.)

d. E. HEFTMANN (editor): Chromatography, Reinhold, New York, 1961. Second edition - 1966.
(Treatise on theory, techniques, and applications. Written by 35 authorities and is a valuable reference for technologists, advanced students, and research workers.)

e. A. I. M. KEULEMANS: Gas Chromatography, Reinhold, New York, 1957; second edition published in 1959.
(An authoritative comprehensive text. Material is clearly and logically presented; classified for easy reference. Good standard reference.)

f. A. B. LITTLEWOOD: Gas Chromatography-Principles, Techniques, and Applications, Academic Press, New York, 1962.
(All aspects of GC are covered.)

g. C. S. G. PHILLIPS: Gas Chromatography, Academic Press, New York, 1956.
(An early work in the field intended to give a broad survey of the work in gas chromatography.)

h. H. PURNELL: Gas Chromatography, J. Wiley & Sons, New York, 1962.
(The practical and theoretical aspects of GC are tied together. Many original ideas and details of technique are presented. Advanced theory.)

i. H. P. BURCHFIELD and E. E. STORRS: Biochemical Applications of Gas Chromatography, Academic Press, New York, 1962.
(Text on GC covering theory and detailed applications to biochemical problems. Covers essential oils, foods, amino acids, carbohydrates, pesticides, clinical chemistry, air pollution, and petroleum chemistry. Indexes included for methods and stationary phases, as well as authors and subjects. Very good, usable applications text.)

j. Compilation of Gas Chromatographic Data, ASTM Special Technical Publication No. 343, 1963.
(Lists liquid phases, solid supports, operating temperature, and relative retention time for over 2, 000 organic compounds. References each entry.)

k. P. G. JEFFERY and P. J. KIPPING: Gas Analysis by Gas Chromatography, MacMillan Co., New York, 1964.
(Discusses gas sampling techniques, proper column selection and detectors used for gas analysis. Suggests methods for determination of common gases, includes gases in solid and liquid samples.)

l. HERBERT H. WOTIZ and STANLEY J. CLARK: Gas Chromatography in the Analysis of Steroid Hormones. Plenum Press, New York, 1966.
(Presents problems of steroid separations, laboratory techniques personally tested by the authors, and extensive bibliography.)

m. HERMAN A. SZYMANSKI: Biomedical Applications of Gas Chromatography, Plenum Press, 1964.
(Collection of reports on applications of G. C. to amines, alkaloids, amino acids, steroids, bile acids, carbohydrates, urinary aromatic acids,

fatty acids and derivatives, and volatile organic anesthetics in blood. References, table of retention times, chromatograms.)

n. MORTIMER B. LIPSETT: Gas Chromatography of Steroids in Biological Fluids, Plenum Press, New York, 1965.
(Presents methods currently (1965) being used for application of G. C. to determination of ketosteroids, corticoids, estrogens and progesterone metabolites. Proceedings of the Workshop on Gas - Liquid Chromatography of Steroids in Biological Fluids, held February 25 - 27, 1965, at Arlie House, Warrenton, Virginia.)

o. L. S. ETTRE, Open Tubular Columns in Gas Chromatography, Plenum Press, New York, 1965.
(Extension bibliography of applicable publications published up to the middle of 1964. Includes theory, operation and preparation of open tubular columns, as well as information on sample introduction, applications, and special techniques.)

p. W. O. McREYNOLDS: Gas Chromatographic Retention Data, Preston Technical Abstracts Co., Evanston, Ill., 1966.
(Lists retention volume and Kovats' Indices calculated at two temperatures for over 300 organic compounds on 79 different liquid phases. Excellent reference book for laboratories doing varied analyses.)

q. ROBERT L. PECSOK: Principles and Practice of Gas Chromatography, Wiley, New York, 1961.
(Compilation of lectures on basic principles of partition chromatography from a non-mathematical point of view. Discussed, also, are proper column selection, sample introduction, and operating procedures.)

r. KEENE P. DIMICK: G. C. Preparative Separations, Varian Aerograph, Walnut Creek, California, 1966.
(A collection of chromatograms, arranged according to the liquid phase used. Includes a brief discussion of the parameters affecting preparative G. C. separations.)

s. A. H. GORDON and J. E. EASTOE: Practical Chromatographic Techniques, George Newnes, London, 1964.
(Directed toward the beginner. Discusses essentials of theory and chromatographic techniques used in paper, adsorption, ion-exchange, and thin-layer chromatography.)

t. EDGAR LEDERER and MICHAEL LEDERER: Chromatography; a review of principles and applications, Elsevier, Amsterdam, 1955.
(A review of chromatographic methods developed within 10-12 years before the publishing date, including adsorption, ion-exchange, and partition chromatography.)

u. HERMAN A. SZYMANSKI, ed: Lectures on Gas Chromatography -- 1962, Plenum Press, New York, 1963.
(Compiled from lectures given at the Advanced Sessions of the Fourth Annual Gas Chromatography Institute, April 23-26, 1962, at Canisius College, Buffalo, New York. Includes G. C. theory, sample introduction, TPGC, and detectors.)

v. HERMAN A. SZYMANSKI and LEONARD R. MATTICK, editors: Lectures on Gas Chromatography-1964: Agricultural and Biological Applications, Plenum Press, New York, 1965.
(Presents a variety of lectures ranging from theoretical to experimental dealing with the indicated

area of interest. Includes determination of pesticides, drugs, blood gases, and fruit volatiles.)

w. AUSTIN V. SIGNEUR, Guide to Gas Chromatography Literature, Plenum Press, New York, 1964. (Consists of "over 7500 references to the published literature and to papers presented.. through late 1963..". Complete author and subject index.)

2. Symposia procedings

a. Vapour Phase Chromatography, edited by D. H. Desty, Butterworths, London, 1957. (Proceedings of the 1st symposium sponsored by the Hydrocarbon Research Group of the Institute of Petroleum, London, 1956.)

b. Gas Chromatography (1957 International Symposium), edited by Vincent J. Coates, et al. Academic Press, New York, 1958. (Proceedings of the Instrument Society of America's 1st International Symposium on Gas Chromatography held at Michigan State University, August, 1957.)

c. Gas Chromatography 1958. Edited by D. H. Desty. Butterworths, London, 1958. (Proceedings of the 2nd Symposium organized by the Gas Chromatography Discussion Group under the auspices of the Hydrocarbon Research Group of the Institute of Petroleum, Amsterdam, 1958.)

d. Gas Chromatography. (1959 International Symposium) Edited by H. J. Noebels, et al. Academic Press, New York, 1961.

(Proceedings of the 2nd International Symposium held under the auspices of the Analysis Instrumentation Division of the Instrument Society of America, June, 1959.)

e. Gas Chromatography 1960. Edited by R. P. W. Scott, Butterworths, Washington, 1960. (Proceedings of the 3rd Symposium organized by the Society for Analytical Chemistry and Gas Chromatography Discussion Group of the Hydrocarbon Research Group of the Institute of Petroleum, Edinburgh, 1960.)

f. Gas Chromatography. (1961 International Symposium) Edited by N. Brenner, et al. Academic Press, New York, 1962. (Proceedings of the 3rd International Symposium held under the auspices of the Analysis Instrumentation Division of the Instrument Society of America, June, 1961.)

g. Gas Chromatography 1962. Edited by M. Van Swaay. Butterworths, Washington, 1963. (Proceedings of the 4th Symposium held under the auspices of the Gas Chromatography Discussion Group of the Hydrocarbon Research Group of the Institute of Petroleum, Hamburg, 1962.)

h. Gas Chromatography (1963 International Symposium). Edited by L. Fowler, Academic Press, New York, 1963. (Proceedings of the 4th International Symposium held under the auspices of the Analysis Instrumentation Division of the Instrument Society of America, June, 1963.)

i. Gas Chromatography 1964. (Reprints of papers, edited by A. Goldup. Institute of Petroleum, London, 1964.)

(Proceedings of the 5th International Symposium, organized by the Gas Chromatography Discussion Group of the Institute of Petroleum, Brighton, September, 1964.)

j. Advances in Gas Chromatography 1965. Edited by A. Zlatkis and L. S. Ettre. Preston Technical Abstracts Co., Evanston, Ill., 1966. (Proceedings of the 3rd International Symposium organized by the University of Houston, Houston, Texas, October, 1965. (Note: Proceedings of 1st and 2nd Symposia published in April 1963 and July 1964 issues of Analytical Chemistry.)

k. Gas Chromatography 1966. Reprints of papers, edited by A. B. Littlewood. Institute of Petroleum, London, 1966. (Proceedings of the 6th International Symposium organized by the Gas Chromatography Discussion Group and held under the auspices of the Hydrocarbon Research Group of the Institute of Petroleum, Rome, September, 1966.)

3. Abstract books of the english gas chromatography discussion group

a. Gas Chromatography Abstracts 1958, edited by: C. E. H. Knapman, Butterworths, London, 1960.

b. Gas Chromatography Abstracts 1959, edited by: C. E. H. Knapman, Butterworths, London, 1960.

c. Gas Chromatography Abstracts 1960, edited by: C. E. H. Knapman, Butterworths, Washington, D. C., 1961.

d. Gas Chromatography Abstracts 1961, edited by: C. E. H. Knapman, Butterworths, Washington, D. C., 1962.

e. Gas Chromatography Abstracts 1962, edited by: C. E. H. Knapman, Butterworths, Washington, D. C., 1963.

f. Gas Chromatography Abstracts 1963, edited by: C. E. H. Knapman, Institute of Petroleum, London, 1964.

g. Gas Chromatography Abstracts 1964, edited by: C. E. H. Knapman, Institute of Petroleum, London, 1965.

h. Gas Chromatography Abstracts 1965, edited by: C. E. H. Knapman, Institute of Petroleum, London, 1966.

i. Gas Chromatography Abstracts 1966, edited by: C. E. H. Knapman, Institute of Petroleum, London, 1967.

(The compilation of these abstracts provides a compact source of valuable information. The books are indexed thoroughly by subject and author.)

4. Periodicals

a. Journal of Chromatography. International Journal on Chromatography, Electrophoresis and Related Methods. (Text in English, French, and German) 1958, monthly. Elsevier Publishing Company, Amsterdam.

b. Journal of Gas Chromatography. International Journal devoted exclusively to various phases of gas chromatography. 1963, monthly, Preston Technical Abstracts Co., Evanston, Ill.

c. Separation Science. Interdisciplinary Journal of Methods and Underlying Processes. 1966, irregular. Marcel Dekker, Inc., New York.

d. Analytical Chemistry. International journal devoted to analytical techniques in chemistry. Monthly. American Chemical Society.

5. Abstract service

GC Abstract Service, Preston Technical Abstract Co., 1718 Sherman Avenue, Evanston, Illinois 60201. (400 journals searched for abstracting. Abstracting done by chemists. Between 30 and 40 abstract issues per week. Most complete source of overall information. Abstracts on 5-inch x 8-inch cards, coded, and punched according to subject matter.)